CW01649266

ITIL® 4 Managing Professional Courseware

Colophon

Title:	ITIL® 4 Managing Professional Courseware
Author:	Van Haren Learning Solutions A.O.
Publisher:	Van Haren Publishing, 's Hertogenbosch
ISBN Hard Copy:	978 94 018 0560 5
Edition:	First edition, first print, October 2019
Design:	Van Haren Publishing, 's-Hertogenbosch
Copyright:	© Van Haren Publishing 2019

For further information about Van Haren Publishing please e-mail us at: info@vanharen.net or visit our website: www.vanharen.net

Publisher about the Courseware

The Courseware was created by experts from the industry who served as the author(s) for this publication. The input for the material is based on existing publications and the experience and expertise of the author(s). The material has been revised by trainers who also have experience working with the material. Close attention was also paid to the key learning points to ensure what needs to be mastered.

The objective of the courseware is to provide maximum support to the trainer and to the student, during his or her training. The material has a modular structure and according to the author(s) has the highest success rate should the student opt for examination. The Courseware is also accredited for this reason, wherever applicable.

In order to satisfy the requirements for accreditation the material must meet certain quality standards. The structure, the use of certain terms, diagrams and references are all part of this accreditation. Additionally, the material must be made available to each student in order to obtain full accreditation. To optimally support the trainer and the participant of the training assignments, practice exams and results are provided with the material.

Direct reference to advised literature is also regularly covered in the sheets so that students can find additional information concerning a particular topic. The decision to leave out notes pages from the Courseware was to encourage students to take notes throughout the material.

Although the courseware is complete, the possibility that the trainer deviates from the structure of the sheets or chooses to not refer to all the sheets or commands does exist. The student always has the possibility to cover these topics and go through them on their own time. It is recommended to follow the structure of the courseware and publications for maximum exam preparation.

The courseware and the recommended literature are the perfect combination to learn and understand the theory.

-- Van Haren Publishing

Other publications by Van Haren Publishing

Van Haren Publishing (VHP) specializes in titles on Best Practices, methods and standards within four domains:

- IT and IT Management
- Architecture (Enterprise and IT)
- Business Management and
- Project Management

Van Haren Publishing is also publishing on behalf of leading organizations and companies: ASLBiSL Foundation, BRMI, CA, Centre Henri Tudor, Gaming Works, IACCM, IAOP, IFDC, Innovation Value Institute, IPMA-NL, ITSqc, NAF, KNVI, PMI-NL, PON, The Open Group, The SOX Institute.

Topics are (per domain):

IT and IT Management
ABC of ICT
ASL®
CATS CM®
CMMI®
COBIT®
e-CF
ISO/IEC 20000
ISO/IEC 27001/27002
ISPL
IT4IT®
IT-CMF™
IT Service CMM
ITIL®
MOF
MSF
SABSA
SAF
SIAM™
TRIM
VeriSM™

Enterprise Architecture
ArchiMate®
GEA®
Novius Architectuur Methode
TOGAF®

Business Management
BABOK® Guide
BiSL® and BiSL® Next
BRMBOK™
BTF
EFQM
eSCM
IACCM
ISA-95
ISO 9000/9001
OPBOK
SixSigma
SOX
SqEME®

Project Management
A4-Projectmanagement
DSDM/Atern
ICB / NCB
ISO 21500
MINCE®
M_o_R®
MSP®
P3O®
PMBOK® Guide
Praxis®
PRINCE2®

For the latest information on VHP publications, visit our website: www.vanharen.net.

Table of content

Self-Reflection of understanding Diagram

'What you do not measure, you cannot control." – Tom Peters

Fill in this diagram to self-evaluate your understanding of the material. This is an evaluation of how well you know the material and how well you understand it. In order to pass the exam successfully you should be aiming to reach the higher end of Level 3. If you really want to become a pro, then you should be aiming for Level 4. Your overall level of understanding will naturally follow the learning curve. So, it's important to keep track of where you are at each point of the training and address any areas of difficulty.

Based on where you are within the Self-Reflection of Understanding diagram you can evaluate the progress of your own training.

Level of Understanding	*Before Training (Pre-knowledge)*	*Training Part 1 (1st Half)*	*Training Part 2 (2nd Half)*	*After studying / reading the book*	*After exercises and the Practice exam*
Level 4 *I can explain the content and apply it .*					
Level 3 *I get it!* *I am right where I am supposed to be.*					Ready for the exam!
Level 2 *I almost have it but could use more practice.*					
Level 1 *I am learning but don't quite get it yet.*					

(Self-Reflection of Understanding Diagram)

Write down the problem areas that you are still having difficulty with so that you can consolidate them yourself, or with your trainer. After you have had a look at these, then you should evaluate to see if you now have a better understanding of where you actually are on the learning curve.

Troubleshooting

Problem areas:	*Topic:*
Part 1	
Part 2	
You have gone through the book and studied.	
You have answered the questions and done the practice exam.	

Timetable

Day 1: ITIL 4 Foundation

- Key concepts, 4 dimensions of SM, ITIL Service Value System, the service value chain & 7 guiding principles

Day 2: Create, Deliver and Support

- How to use Value streams with support of ITIL practices

Day 3: Drive Stakeholder Value

- Customer journeys, relationships and service offerings
- How to act together to ensure continual value co-creation

Day 4: High Velocity IT

- The Digital enterprise and how to use guiding principles

Day 5: Deliver, Plan and Improve and Exam

- Direction and Planning, GRC, Communication & OCM
- Work with sample exam and examination

Welcome to ITIL 4 Managing Professional Transition

ITIL®
ACCREDITED TRAINING ORGANIZATION
PeopleCert ON BEHALF OF AXELOS

COURSEWARE

Van Haren PUBLISHING

Our coming days

- ✓ Safety
- ✓ Breaks
- ✓ Lunch
- ✓ Phones/Computers

Our coming days
Active participation!
Ask questions
Theory (a lot)
Some homework
Group dialogues
Copyright © Van Haren Publishing and AXELOS Limited This material is based on AXELOS ITIL® material.
Material is reproduced under license from AXELOS. All rights reserved
3

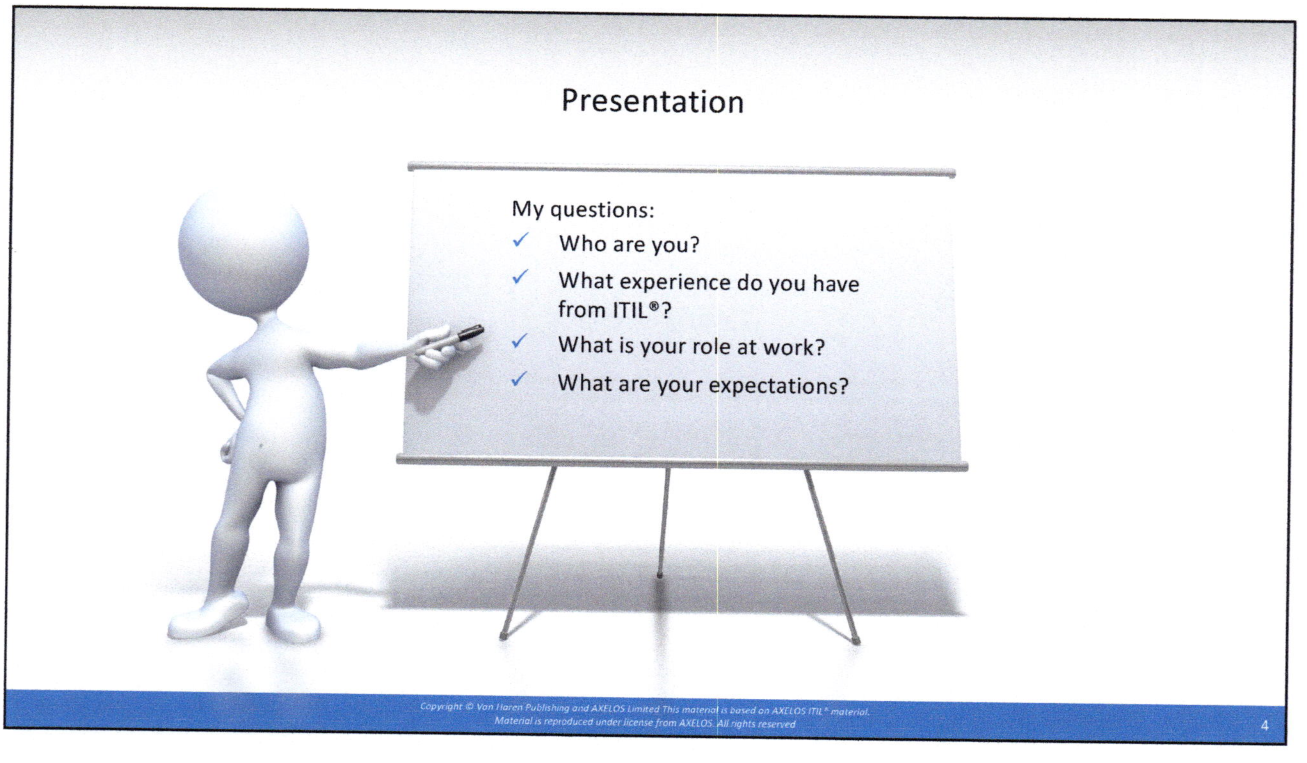

Presentation
My questions:
Who are you?
What experience do you have from ITIL®?
What is your role at work?
What are your expectations?
Copyright © Van Haren Publishing and AXELOS Limited This material is based on AXELOS ITIL® material.
Material is reproduced under license from AXELOS. All rights reserved
4

Course schedule

- ✓ Day 1: ITIL 4 Foundation
  - Key concepts, 4 dimensions of SM, ITIL Service Value System, the service value chain & 7 guiding principles
- ✓ Day 2: Create, Deliver and Support
  - How to use Value streams with support of ITIL practices
- ✓ Day 3: Drive Stakeholder Value
  - Customer journeys, relationships and service offerings
  - How to act together to ensure continual value co-creation
- ✓ Day 4: High Velocity IT
  - The Digital enterprise and how to use guiding principles
- ✓ Day 5: Deliver, Plan and Improve and Exam
  - Direction and Planning, GRC, Communication & OCM
  - Work with sample exam and examination

ITIL 4: INTRODUCTION

- Some key drivers behind this new version of ITIL

ITIL is evolving…

- From process focus to holistic view
- From fragmented lifecycle to end-to-end
- From major releases to continual improvement
- From operational silos to flexible value flow

- Since the latest update of ITIL in 2011 we have seen huge changes impacting most organizations who have adopted ITIL best practices. There is a clear need to adapt to changes in markets, technologies and ways of working and ITIL is no exception.

Principles as a central theme

- There has been a clear trend among the commonly used frameworks, models and methodologies to move away from rules and focus more on principles.

- We see this development as a positive thing in general as it can make interoperability and integration between frameworks and methodologies more accepted as most basic underpinning principles often are shared.

Product management

- There is a trend among IT organizations to use a product management terminology and adopt ways of working from other domains like engineering or manufacturing. Organizations own or have access to a variety of resources (and capabilities). Products are configurations of these resources that will potentially be valuable for its customers. The management of these products has become synonymous with service management.

- This is not new as it was already partly introduced in previous versions of ITIL but makes ITIL 4 even more relevant and easier to relate to the emerging agile movement with its strong emphasis on development and management of products.

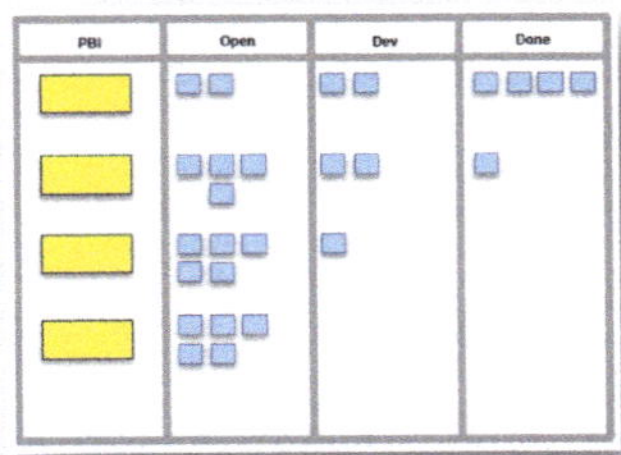

ITIL 4:
MP TRANSITION

- Enables ITIL v3 practitioners to reach the designation of ITIL 4 Managing Professional

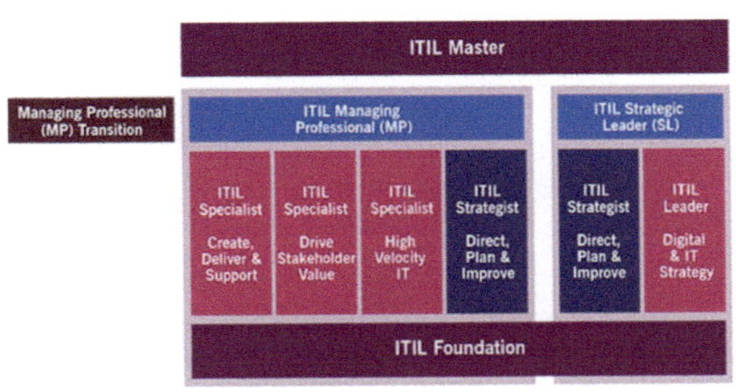

Not a regular ITIL 4 publication...

- There is no separate ITIL 4 publication for this course. Content originates from:
- Foundation
- Create, Deliver and Support
- Drive Stakeholder Value
- High Velocity IT
- Direct, Plan and Improve

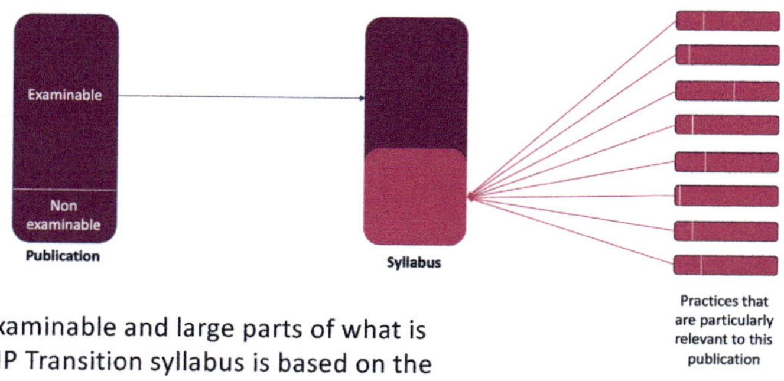

- Not all content in the publications is examinable and large parts of what is examinable relates to practices. The MP Transition syllabus is based on the other separate ITIL 4 syllabi and exams. Mainly focusing on new areas and with very little content related to practices

A slightly different exam...

- The exam is a closed book exam that consists of 40 multiple-choice questions with four options (same format as an ITIL Foundation exam)

  The pass rate is set at 70% (28 correct answers) and the available time is 90 minutes. Non-English-speaking candidates an additional 25% extra time is awarded (113 minutes)

- Most questions are at Blooms taxonomy level 3 that require the candidate to demonstrate that they know how to apply the concepts. The rest are at Bloom level 2 that require the candidate to be able to explain, describe and understand the concepts

- This means that there are no questions at higher Bloom levels that require the candidate to do an analysis or synthesis of the concepts

That give you a special route…

- This course provides experienced practitioners with the ITIL 4 Managing Professional designation, although not fully certifying the same level of knowledge as taking all modules separately
-

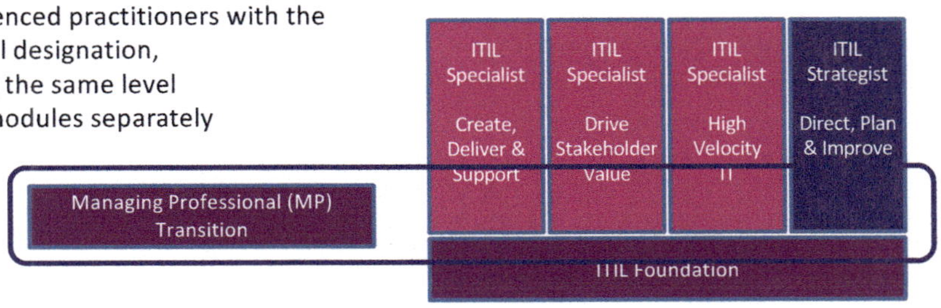

The MP Transition course covers only 30-40% of what is in the separate ITIL 4 syllabi and exams. There will be separate publications covering all 34 practices

Additional study of all these publications and practices is therefore advised!

ITIL 4: FOUNDATION

- Understand the key concepts of service management

Key learning requirements

FOUNDATION

- Understand the key concepts of service management
- Understand how the ITIL guiding principles can help an organization adopt and adapt service management
- Understand the four dimensions of service management
- Understand the purpose and components of the ITIL service value system
- Understand the activities of the service value chain, and how they interconnect

Introduction

Most organizations need to address service management challenges and utilize the potential of modern technology. ITIL 4 is designed to ensure a flexible, coordinated and integrated system for the effective governance and management of IT-enabled services.

Key terms defined and important concepts of service management introduced here include:

- organizations, service providers, service consumers, and other stakeholders
- products and services
- the nature of value and value co-creation
- service relationships
- value: outcomes, costs, and risks.

These concepts ***apply to all*** *organizations and services, regardless of their nature and underpinning technology.*

What is service management?

Definition: Service management
A set of specialized organizational capabilities for enabling value for customers in the form of services.

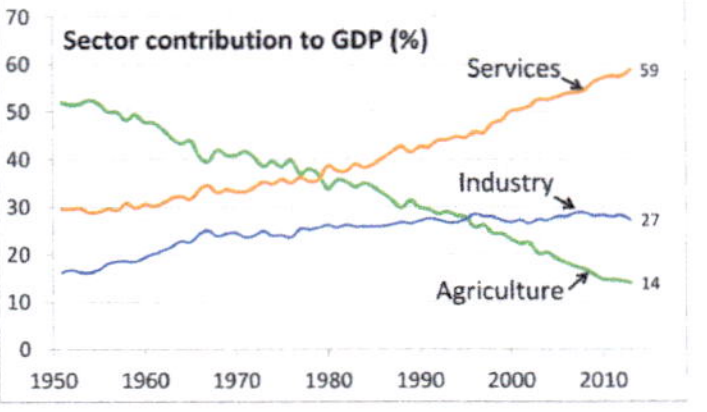

Developing these capabilities requires an understanding of:

- the nature of value
- the nature and scope of the stakeholders involved
- how value creation is enabled through services.

Value creation

The **purpose of an organization** is to **create value** for stakeholders.

The term **'value'** is commonly used in service management, and it is a key focus of ITIL 4; it must therefore be clearly defined.

Definition: Value
Value is the perceived benefits, usefulness and importance of something

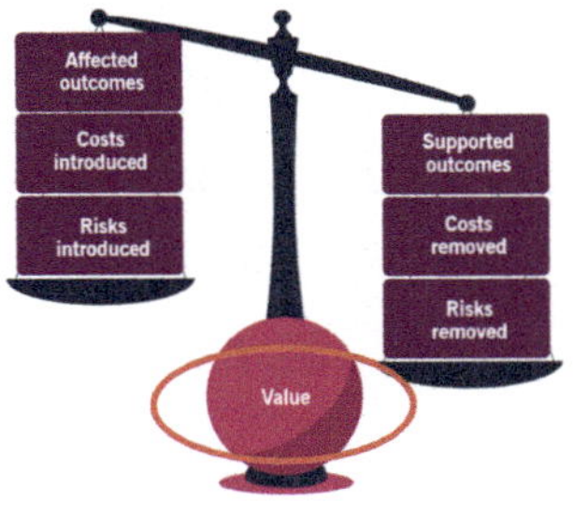

Value creation is a balancing act involving **outcomes**, **costs** and **risks.**

Note: Value can be subjective and therefore different according to each stakeholder!

Value co-creation is essential

The bad example: "Service provider" acting in isolation

The provider delivers the service and the consumer receives value, but the **consumer plays no role** in the creation of value for themselves.

The good example: Active collaboration between provider and consumer

Increasingly more consideration is put into the highly **complex** and **interdependent** service **relationships** that exist in reality. **Both** parties **actively** seek to establish **mutually beneficial**, interactive relationships with their consumers, empowering them to be creative collaborators in the **service value chain**.

Stakeholders across the service value chain contribute to the definition of requirements, design of service solutions and even to service creation and/or provisioning itself.

Service providers and consumers

Service providers:

Can be **external** or **internal.** Could be selling services on the **open market** to **other businesses**, or to **individual consumers.**

Service providers can be **part of** a service alliance, collaborating to provide services to consumer organizations.

Service consumers:

When **receiving services**, an organization takes on the role of the service consumer.

It is a **generic role** that is used to simplify, in practice, more roles involved such as **customers**, **users** and **sponsors**. These roles can be **combined**.

*Clear understanding of **who its consumers are** in a given situation and who the other stakeholders are in the associated **service relationships**. These roles may have different expectations from services, and **different definitions of value**.*

Key stakeholders

Definition: Organization

A person or a group of people that has its own functions with responsibilities, authorities, and relationships to achieve its objectives.

Definition: Customer

A person who defines the requirements for a service and takes responsibility for the outcomes of service consumption.

Definition: User

A person who uses services.

Definition: Sponsor

A person who authorizes budget for service consumption.

Other stakeholders and value

- Beyond the consumer and provider roles, there are many other stakeholders that are important to value creation.

Stakeholder	Example of value for stakeholder
Service consumers	Benefits achieved; costs and risks optimized
Service provider	Funding from the consumer; business development; image improvement
Service provider employees	Financial and non-financial incentives; career and professional development; sense of purpose
Society and community	Employment; taxes; organizations' contribution to the development of the community
Charity organizations	Financial and non-financial contributions from other organizations
Shareholders	Financial benefits, such as dividends; sense of assurance and stability

- It is important that relationships with all key stakeholders are understood and managed.

Services and products

The central component of service management is, of course, the service.

Definition: Service

A means of enabling value co-creation by facilitating outcomes that customers want to achieve, without the customer having to manage specific costs and risks.

The services that an organization provides are based on one or more of its products. Organizations own or have access to a variety of resources. Products are configurations of these resources, created by the organization, that will potentially be valuable for its customers.

Definition: Product

A configuration of an organization's resources designed to offer value for a consumer.

Services offerings

- Service providers present their services to consumers in the form of service offerings, which describe one or more services based on one or more products.

Definition: Service offering

A description of one or more services, designed to address the needs of a target consumer group. A service offering may include goods, access to resources, and service actions.

Component	Description	Examples
Goods	Supplied to the consumer Ownership is transferred to the consumer Consumer takes responsibility for future use	A mobile phone A physical server
Access to resources	Ownership is not transferred to the consumer Access is granted or licensed to the consumer under agreed terms and conditions The consumer can only access the resources during the agreed consumption period and according to other agreed service terms	Access to the mobile network, or to network storage
Service actions	Performed by the service provider to address a consumer's needs Performed according to an agreement with the consumer	User support Replacement of a piece of equipment

Service relationships

A **key focus** of service management, and of ITIL, is the way that organizations **co-create value** with their consumers through **service relationships**. Service relationships are established between two or more organizations to create value.

The roles of service provider and service consumer are not mutually exclusive, and organizations typically both provide and consume a number of services at any given time.

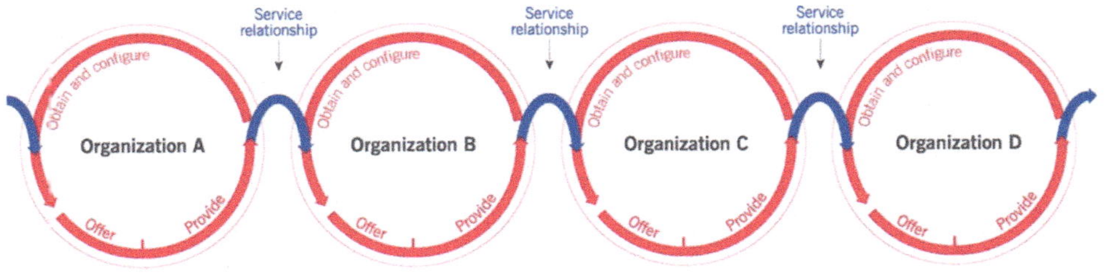

Service relationships

Definition: Service relationship management consists of **joint activities** performed by a service provider and a service consumer **to ensure continual value co-creation** based on agreed and available service offerings.

Definition: Service provisioning consists of activities performed by a service provider to provide services. This includes provision of access to resources, fulfillment of service actions, resource management and service performance and continual improvement

Definition: Service consumption consists of activities performed by a service consumer to consume services. This includes the management of the consumers resources, utilization of providers resources, requesting of service actions and acquiring of goods

Differentiate output and outcome

It is important to differentiate between output and outcome of services.

Definition: Output
A tangible or intangible deliverable of an activity.

Definition: Outcome
A result for a stakeholder enabled by one or more outputs.

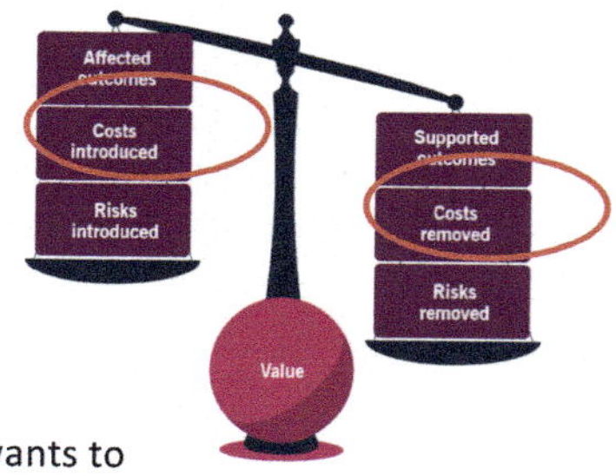

be difficult for the provider **to fully understand the outcomes** that the consumer wants to achieve.

*Some service providers **predict** or even **create demand** for certain outcomes, forming a target group for their services.*

Two types of costs to understand

Definition: Cost
The amount of money spent on a specific activity or resource.

From the service consumers perspective there are **two types**:
Costs removed from the consumer by the service (a part of the value proposition).

Costs introduced or imposed on the consumer by the service (the costs of service consumption). Some consumers describe this as what they have to 'invest' to consume the service.

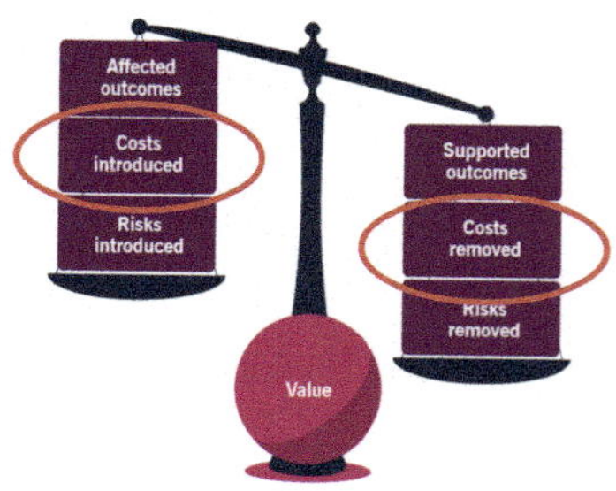

Balanced approach needed to cost for both parties in the service relationship.

Two types of risk to understand

Definition: Risk

A possible **event** that could cause **harm** or **loss** or make it more difficult to achieve objectives. Can also be defined as "**uncertainty of outcome**".

From the service consumers perspective there are **two types**:

Risks removed from the consumer by the service (a part of the value proposition).

Risks introduced or imposed on the consumer by the service (risks of service consumption).

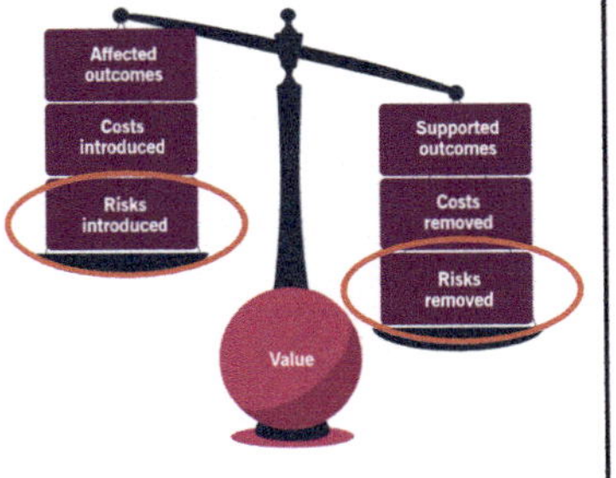

Balanced approach needed to risk reduction throughout the service relationship.

Definitions of utility and warranty

Utility:

- The functionality offered by a product or service to meet a particular need.
- Utility can be summarized as 'what the service does' and can be used to determine whether a service is 'fit for purpose'.
- To have utility, a service must either support the performance of the consumer or remove constraints from the consumer. Many services do both.

Warranty:

- Assurance that a product or service will meet agreed requirements.
- How the service performs and can be used to determine whether a service is 'fit for use'.
- Often a formal agreement, related to service levels, or a marketing message or brand image. A 'warranty', if all defined and agreed conditions are met in areas such as availability, capacity, security and continuity levels of service.

***Both** are essential for a service to facilitate its desired outcomes and create value.*

Summary

We have just talked about:

- ✓ This section has covered the key concepts in service management, in particular the nature of value and value co-creation, organizations, products and services
- ✓ It has explored the often complex relationships between service providers and consumers, and the various stakeholders involved.
- ✓ The chapter has also covered the key components of consumer value: outcomes, costs and risks, and how important it is to understand the needs of the customer when designing and delivering services
- ✓ These concepts will be built upon throughout this training, and guidance provided on applying them in practical and flexible ways
- ✓ After completing this section you should be able to recall several definitions, understand and be able to describe several key concepts.

Q: Identify the missing word(s) in the following sentence

A service is a means of enabling value co-creation by facilitating [?] that customers want to achieve.

A. the warranty

B. outcomes

C. the utility

D. outputs

Q: Which is the definition of warranty?

A. A tangible or intangible deliverable that is produced by carrying out an activity

B. The assurance that a product or service will meet agreed requirements

C. A possible event that could cause harm or loss, or make it more difficult to achieve objectives

D. The functionality offered by a product or service to meet a particular need

Q: A service provider describes a package that includes a laptop with software, licenses, and support.

What is this package an example of?

A. Value

B. An outcome

C. Warranty of a service

D. A service offering

Q: What are the two types of cost that a service consumer should evaluate?

A. The cost of creating the service, and the cost charged for the service

B. The costs removed by the service, and the costs imposed by the service

C. The cost of provisioning the service, and the cost of improving the service

D. The cost of purchasing software, and the cost of purchasing hardware

FOUR DIMENSIONS

Understand the four dimensions of service management

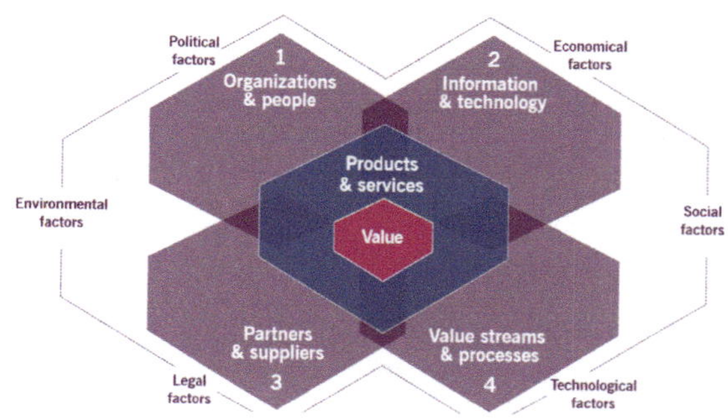

Four dimensions

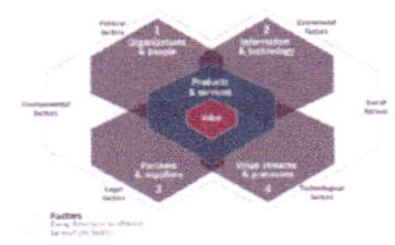

- Previously referred to as "the 4 P's of ITSM" these dimensions are well used in ITIL. They are still relevant to, and impact upon, all practices, the entire service value chain and the whole Service Value System (SVS).

- These four dimensions are:
- Organizations and people
- Information and technology
- Partners and suppliers
- Value streams and processes

*Note: The four dimensions do **not** have **sharp boundaries** and **may overlap,** and apply to **all services** being managed!*

- Organizations should holistically consider all aspects of their behavior, but will often become too focused on one dimension in their initiatives and neglect others.

1. Organizations and people

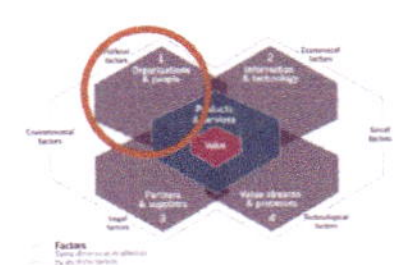

The first dimension of service management – **organizations & people** – is important to ensure that the way an organization is **structured** and **managed**, as well as **its roles, responsibilities, systems of authority** and **communication** are well-defined and **support** its over-all **strategy** and **operating** model.

Other important aspects of this dimension are:

- A **culture** that supports its **objectives**, **capacity** and **competence** among its workforce
- Leaders advocate **values** that **motivate people** – the **key element**
- Understanding the **interfaces**, to ensure proper levels of **collaboration** and **coordination**
- **Everyone** should have a focus on **value creation** and **broad general knowledge**

Adopting the ITIL guiding principles can be a good starting point

2. Information and technology

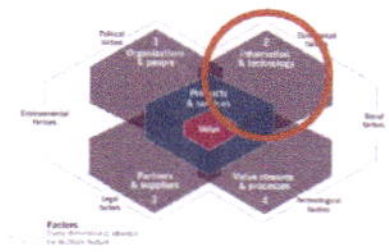

The technologies that support service management include, but are not limited to, workflow management systems, knowledge bases, inventory systems, communication systems and analytical tools.

In the context of a **specific IT service**, this dimension includes the information created, managed and used in the course of service provision and consumption, and the technologies that support and enable that service.

The IT architecture, including applications, databases, communication systems and their integrations and the use of technology is a business differentiator.

The culture of an organization may have significant impact on the technologies it chooses

Information management

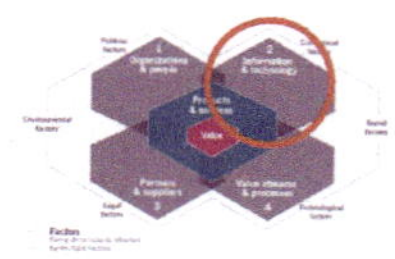

Information management is a means of enabling business value and **information** is generally the **key output** of a majority of IT services provided

Another **key consideration** in this dimension is **how information is exchanged** between different services and service components

The information architecture needs to be **well-understood** and **continually optimized** – taking into account such **information criteria** as; **availability**, **reliability**, **accessibility**, **timeliness**, **accuracy**, and **relevance** of the information provided

Focus for this dimension are also **security**, **regulatory** and **compliance requirements**

Technology - good questions to ask

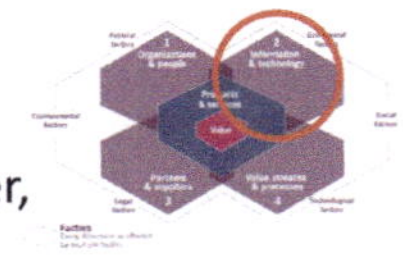

- Is this technology **compatible with the current architecture** of the organization and its customer(s)?
- Do the different technology products used **work together**?
- How are emerging technologies (such as machine learning, artificial intelligence and Internet of Things) **likely to disrupt** the service or the organization?
- Does this technology **raise any regulatory or other compliance issues** with the organization's policies and information security controls, or those of its customers?
- Is this a technology that will **continue to be viable** in the foreseeable future?
- Is the organization willing to accept the risk of using aging technology, or of embracing emerging or unproven technology?
- Does this technology **align with the strategy** of the provider, or its service consumers?
- Does the organization **have the right skills across its staff and suppliers** to support and maintain the technology?
- Does this technology **have sufficient automation capabilities** to ensure it can be efficiently developed, deployed and operated?
- Does this technology **offer additional capabilities** that might be leveraged for other products or services?
- Does this technology **introduce new risks or constraints** to the organization (for example, locking it into a specific vendor)?

ITSM - cloud computing

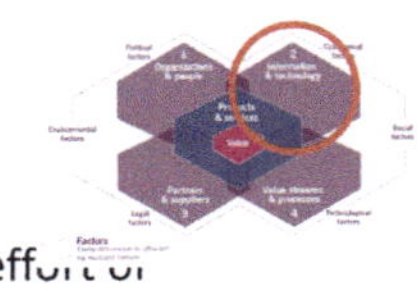

Definition: Cloud computing

A model for enabling on-demand network access to a shared pool of configurable computing resources that can be rapidly provided with minimal management effort or provider interaction.

Cloud computing **changes** service architecture and distribution of **responsibilities** between service consumers, service providers and their partners. It may enable significantly **faster deployment** of new and changed services, thus supporting **high velocity** service delivery.

Considering the influence of cloud computing on organizations, it is important to make decisions about the use of this model at the strategic level of the organization, **involving all levels of stakeholders**, from governance to operations.

Cloud computing impact on ITSM

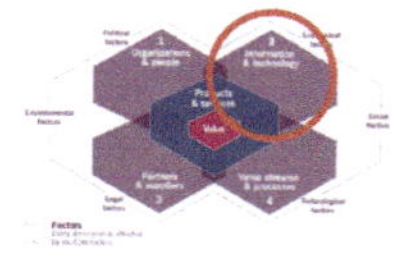

Replaces some infrastructure, previously managed by the service provider, with a partner's cloud service.

Decreases or removes the need for infrastructure management expertise and resources.

Shifts focus of service monitoring and control from in-house infrastructure to the cloud.

Changes cost structures of the service provider, removing specific capital expenditures and introducing new operating expenditures.

Introduces higher requirements for network availability and security.

Introduces new security and compliance risks and requirements, applicable to both the service provider and their partner providing the cloud service.

Provides users with opportunities to scale service consumption using self-service, via simple standard requests, or even without any requests.

Cloud computing affects practices

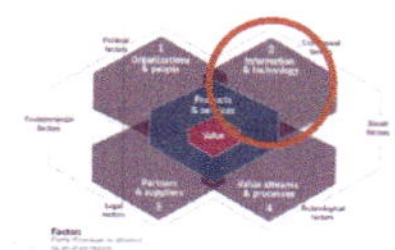

Practices that are affected by cloud computing, include, but are not limited to:

- service level management
- measurement and reporting
- information security management
- service continuity management
- supplier management
- incident management
- problem management
- service request management
- service configuration management

3. Partners and supplier

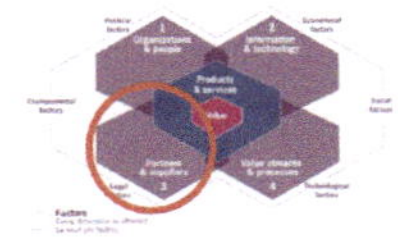

Every organization and every service **depends,** to some extent, on **services provided by other organizations.**

This dimension covers relationships with other organizations that are **involved** in the **design**, **development**, **deployment**, **delivery**, **support** and/or **continual improvement of services.**

Contracts and agreements - This ranges from formal contracts with clear separation of responsibilities, to flexible partnerships where parties share common goals and risks, and collaborate to achieve desired outcomes.

Form of cooperation	Outputs	Responsibility for the outputs	Responsibility for achievement of the outcomes	Level of formality	Examples
Goods supply	Goods supplied	Supplier	Customer	Formal supply contract/invoices	Procurement of computers and phones
Service delivery	Services delivered	Provider	Customer	Formal agreements and flexible cases	Cloud computing (infrastructure of platform as a service)
Service partnership	Value co-created	Shared between provider and customer	Shared between provider and customer	Shared goals, generic agreements, flexible case-based arrangements	Employee onboarding (shared between HR, facilities and IT)

Different strategies for cooperation

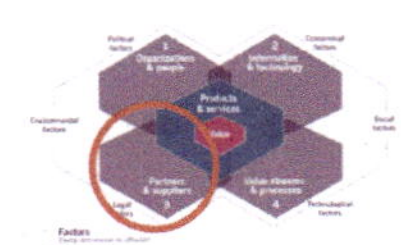

Forms of **cooperation** depends on strategy and objectives for customer relationships. **Strategy** to using partners and suppliers should be based on its **goals**, **culture** and **business environment**. Influencing factors include:

Strategic focus - to focus on core competencies or to stay as self-sufficient as possible
Corporate culture - long-standing cultural bias is difficult to change
Resource scarcity - resources or skillsets in short supply are acquired from suppliers
Cost concerns - more economical to source a particular requirement from a supplier
Subject matter expertise - less risky to use a supplier that already has an expertise
External constrains - regulation or policy constraints might impact a strategy
Demand patterns - customer activity or demand for services might be seasonal or variable

Service Integration and Management

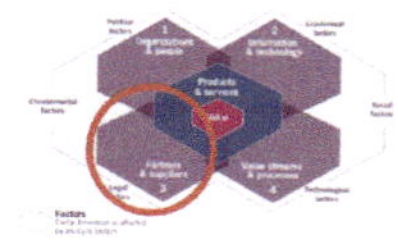

One method an organization may use the partners and suppliers dimension is **service integration and management (SIAM)**. This uses an integrator to help **ensure** that **service relationships** are **properly coordinated.**

Service integration and management **may be kept inside** the organization, but can also be **delegated** to a **trusted partner**.

> ***'as a service'*** *– a bundle of goods and services to a single product that can be consumed as a utility*

4. Value streams and processes

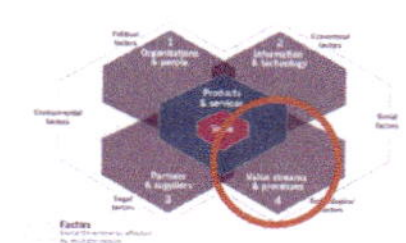

The value streams and processes dimension applies **to both** the **Service Value System (SVS)** in general, and to specific **products** and **services**.
In **both contexts** it defines activities, workflows, controls and procedures needed to achieve agreed objectives. The dimension **focuses on** what **activities** the organization undertakes, and how they are **organized**, as well as how the organization ensures that it is **enabling value creation** for all stakeholders **efficiently** and **effectively**.

ITIL gives service providers an **operating model** that **covers all** the **key activities** required to effectively manage products and services, called the **ITIL service value chain**.

The service value chain operating model is **generic**; however, in practice it can **follow different patterns**. These **patterns** within the value chain operation are **called value streams**.

Value streams

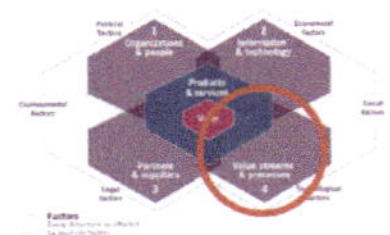

Definition: Value stream

A **series of steps** an organization undertakes to create and deliver products and services to consumers.

A value stream is a **combination of** the organization's **value chain activities**. Identifying and understanding the organizations value streams is critical to improving its performance. Structuring its service and product portfolios around value streams gives organizations a **clear picture** of what they deliver and how, and how to continually improve.

Value stream optimization may include **process automation** or adoption of emerging technologies and ways of working to gain efficiencies or enhance user experience and eliminate waste. Value streams should be **adoptable** and **continually improved**.

Processes

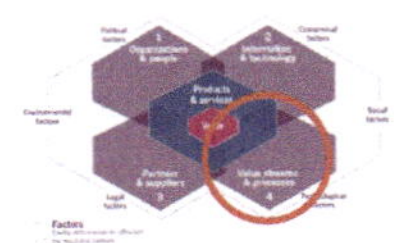

Definition: Process

A set of interrelated or interacting **activities** that transform inputs into outputs. A process takes one or more **defined inputs** and turns them **into defined outputs**. Processes define the **sequence** of actions and their dependencies.

Processes **describe** what is **done to accomplish an objective**, and well-defined processes can improve productivity within and across organizations. They are usually **detailed** in **procedures**, which outline who is involved in the process, and **work instructions**, which explain how they are carried out.

The same structure (of the value chain, value streams, processes, procedures and work instructions) applies to services: to successfully **create, deliver** and **improve a service.**

External factors

Other factors that influence the four dimensions are:

- **P**olitical factors
- **E**conomical factors
- **S**ocial factors
- **T**echnological factors
- **L**egal factors
- **E**nvironmental factors

Collectively, these factors influence how organizations configure their resources and address the four dimensions of service management.

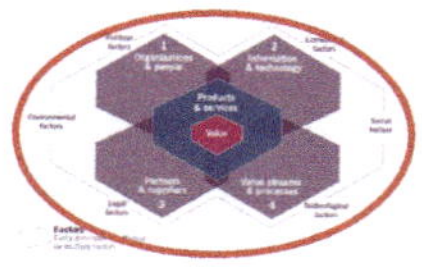

Political	Economic	Social
• Tax policies • Fiscal policy • Trade tariffs • Change of government • Local government policy (eg planning consents)	• Inflation rate • Interest rates • Foreign exchange rates, • Economic growth patterns	• Cultural trends • Demographics • Employee expectations • Population analytics • Buying trends • Seasonal behaviors
Technological	**Legal**	**Environmental**
• Automation • Research and development • Technical awareness in the market • Impact of new media	• Consumer laws • Health and safety standards, • Labour laws etc. • Trade barriers	• Geographical location • Climate change • Environmental offsets • Emissions legislation • Green agenda

Summary

We have just talked about:

- ✓ This section has covered the four dimensions of ITSM:
  1. Organizations and people
  2. Technology and information
  3. Partners and suppliers
  4. Value streams and processes
- ✓ Other external factors to consider include:
  - Political factors
  - Economical factors
  - Social factors
  - Technological factors
  - Legal factors
  - Environmental factors
- ✓ Every dimension is affected by multiple factors and influence how service providers operate

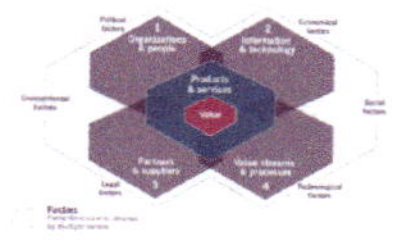

Q: Which service management dimension is focused on activities and how these are coordinated?

A. Organizations and people

B. Information and technology

C. Partners and suppliers

D. Value streams and processes

Q: Which is NOT a key focus of the 'information and technology' dimension?

A. Security and compliance

B. Communication systems and knowledge bases

C. Workflow management and inventory systems

D. Roles and responsibilities

Service value system

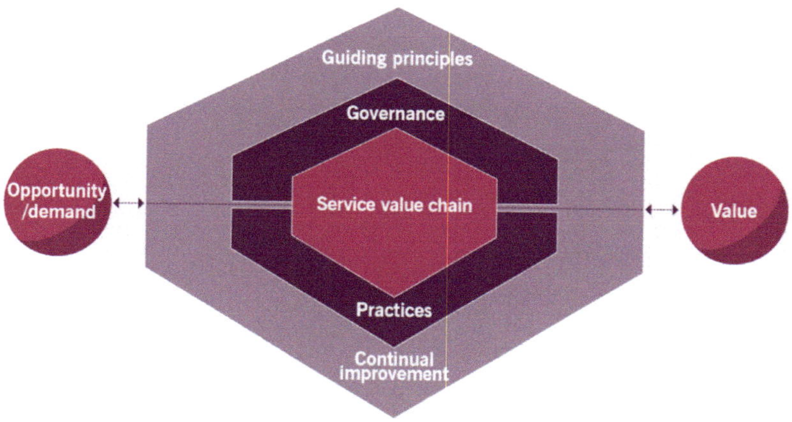

ITIL service value system (SVS)

To enable value creation, **all** the **components and activities** of the organization need to **work together as a system**. The three main parts of the ITIL SVS are:

- **Inputs** to the system:
  - **Opportunity** and **demand**
- **Components** of the system:
  - **Guiding principles**
  - **Governance**
  - **Service value chain**
  - **Practices**
  - **Continual improvement**
- **Outputs** of the system:
  - Achievement of **objectives** and **value**

Service value system

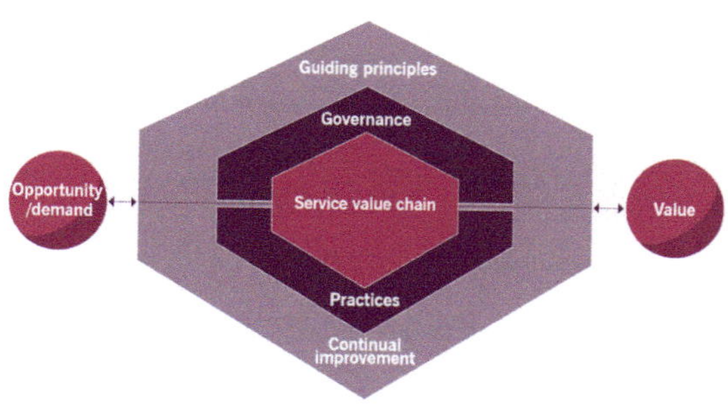

Purpose of the SVS

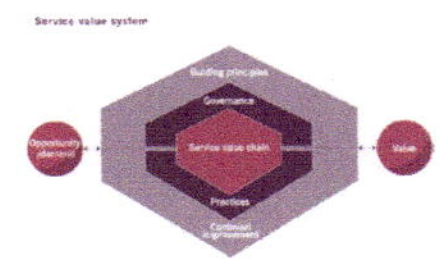

Purpose:

- To ensure that the organization continually co-creates value with all stakeholders through the use and management of products and services.

The ITIL Service Value System describes how all the components and activities of the organization work together as a system to enable value creation.

These activities represent the steps an organization takes in the creation of value. Each activity may draw upon internal or third-party resources, processes, skills, and competencies from one or more practices.

Inputs and outputs of the SVS

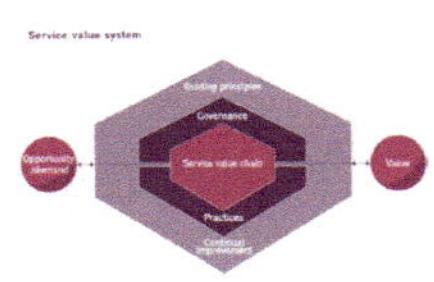

Each organization's SVS has interfaces with other organizations, forming an **ecosystem**. The inputs and outputs of the SVS can be described further as:

Inputs:

- **Opportunity** – represents options or possibilities to add value for stakeholders or otherwise improve the organization
- **Demand** – need or desire for products and services among internal and external consumers

Outcome:

- **Value** – the perceived benefits, usefulness and importance of something

ITIL SVS can enable creation of different types of value for a wide group of stakeholders.

Components of the SVS

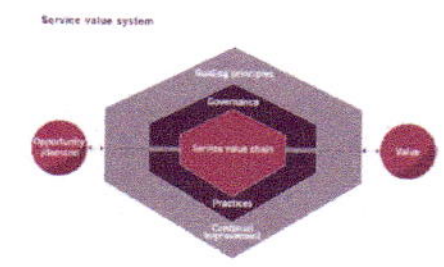

Guiding principles

- Recommendations that can guide an organization in all circumstances, regardless of changes in its goals, strategies, type of work, or management structure.

Governance

- The means by which an organization is directed and controlled.

Service value chain

- A set of interconnected activities that an organization performs to deliver a valuable product or service to its consumers and to facilitate value realization.

Practices

- Sets of organizational resources designed for performing work or accomplishing an objective.

Continual improvement

- A recurring organizational activity performed at all levels to ensure that an organization's performance continually meets stakeholders' expectations.

Challenges

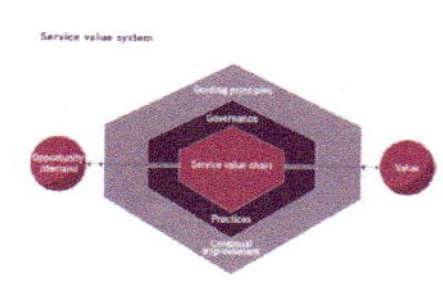

Components and activities, together with the organization's resources, can be **configured** and **reconfigured** in **multiple combinations** in a **flexible** way as circumstances change. But this requires the **integration** and **coordination** of activities, practices, teams, authorities and responsibilities and all parties to be truly effective.

Challenges:

- **Organizational silos** makes it hard to work effectively and efficiently with a shared vision, or to become more agile and resilient
- **Silos can be resistant to change** and **prevent easy access to the information** and **specialized expertise**, which can reduce efficiency and increase cost and risk
- Silos also make **communication** or **collaboration** more difficult across groups.

Challenges addressed with SVS

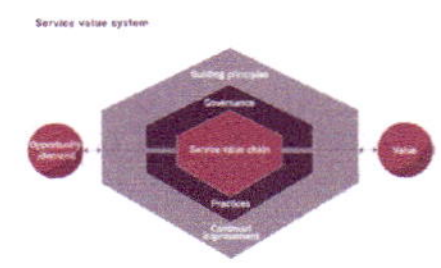

The ITIL SVS has been specifically **architected** to **enable flexibility** and **discourage** siloed working. It does not form a **fixed**, rigid **structure**. There are examples of value streams, but they are **not definite or prescriptive**.

Organizations should be able to define and redefine their value streams in a **flexible, yet safe** and **efficient** manner. This requires **continual improvement** activity to be carried out at all levels of the organization.

The **guiding principles create a foundation** for a shared culture across the organization. ITIL SVS **supports many work approaches**, such as Agile, DevOps and Lean, as well as traditional process and project management, with a **flexible value-oriented operating model**.

Summary

We have just talked about:

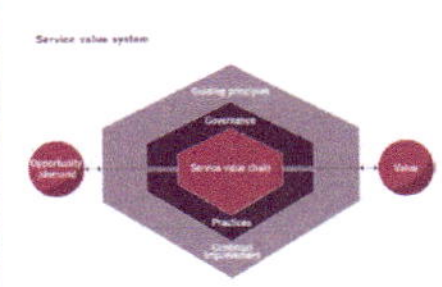

- ✓ This section has covered the ITIL Service Value System (SVS) and its purpose, function, structure and business value
- ✓ The three main parts of the SVS and their content:

  Inputs:
  - *Opportunity* and *demand*

  Components:
  - *Guiding principles*
  - *Governance*
  - *Service value chain*
  - *Practices*
  - *Continual improvement*

  Outcome:
  - *Value to stakeholders*
- ✓ The reason for the architecture of the SVS and how it addresses challenges with a siloed approach

Q: Which ITIL concept describes governance?

A. The seven guiding principles

B. The four dimensions of service management

C. The service value chain

D. The service value system

THE SERVICE VALUE CHAIN

- Understand the activities of the service value chain, and how they interconnect

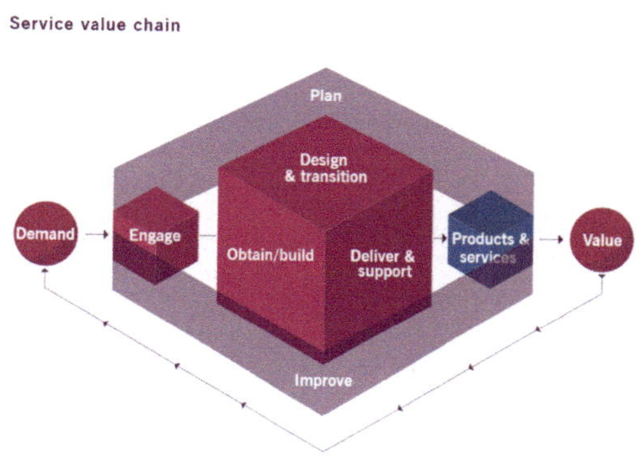

Copyright © Van Haren Publishing and AXELOS Limited This material is based on AXELOS ITIL® material.
Material is reproduced under license from AXELOS. All rights reserved

A central element of the SVS

The **service value chain**, an operating model which **outlines** the **key activities** required to **respond** to **demand** and **facilitate value realization** through the **creation** and **management** of **products** and **services**. Its **six key activities** are:

- **Plan**
- **Improve**
- **Engage**
- **Design & transition**
- **Obtain/build**
- **Deliver & support.**

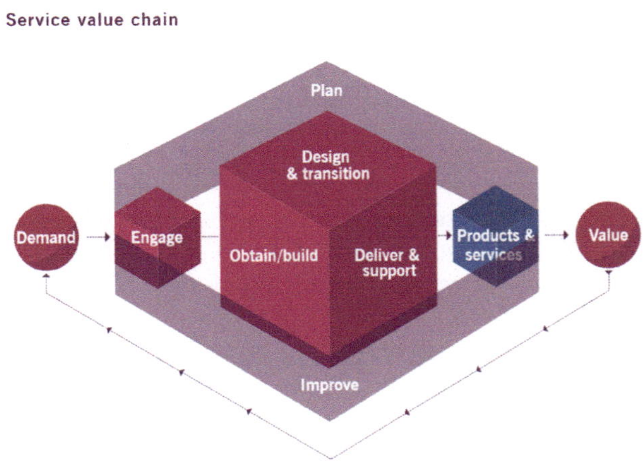

The service value chain

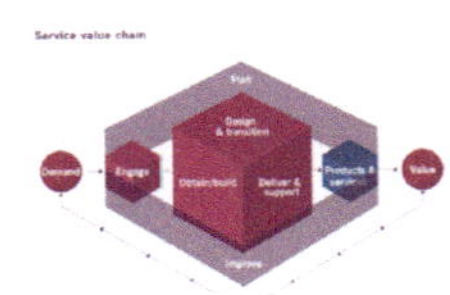

The **activities** are **connected to**, and **interact with** one another, with each activity **receiving** and **providing triggers** for further **actions** to be taken.

To convert **inputs** into **outputs**, the value chain activities use different combinations of **ITIL practices** (sets of resources designed for performing certain types of work).

For example:
*The **engage** value chain **activity** might **draw upon** a number of **practices** including **supplier management**, **service desk management**, **relationship management** and **service request management** to **respond to** new **demands** for products and services, decisions, or information **from various stakeholders**.*

The service value chain

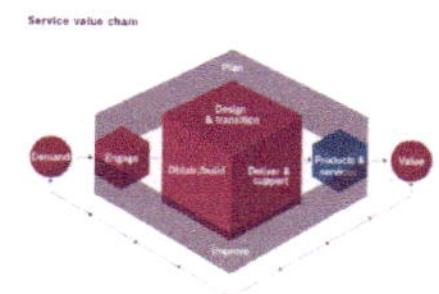

Different activities have different focuses:

- All incoming and outgoing interactions with parties external to the service provider are performed via the **engage** value chain activity
- All **new** resources or capabilities are obtained through the **obtain/build** activity
- Planning at all levels is performed via the **plan** value chain activity
- Improvements at all levels are initiated and managed via the **improve** activity
- Creation, modification, delivery, maintenance and support of components, products and services are performed in an integrated and coordinated way between the **design and support**, **obtain/build** and **deliver and support** activities

*Note: **Products & Services, Demand** and **Value** are not considered value chain activities!*

Service value streams

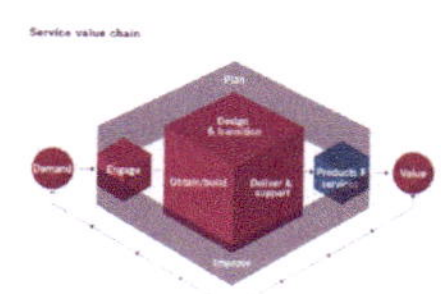

Service value streams are specific **combinations** of **activities** and **practices**, and each one is **designed for a particular scenario**. Once designed, value streams should be subject to **continual improvement**.

Example of generic **practices** that can be used to support many different scenarios:

- Business analysis
- Development
- Testing
- Release and deployment
- Support.

Note: Although the high-level steps are universal, different products and clients need different streams of work.

These practices are **applied differently**, depending on scenario and product or service, in **different value streams** and **combine different practices** and **value chain activities**.

Plan

The purpose of this value chain activity is to ensure a **shared understanding** of the **vision**, current **status** and improvement **direction** for all four dimensions and all products and services across the organization

Note: Planning at **all levels** is performed via the plan activity

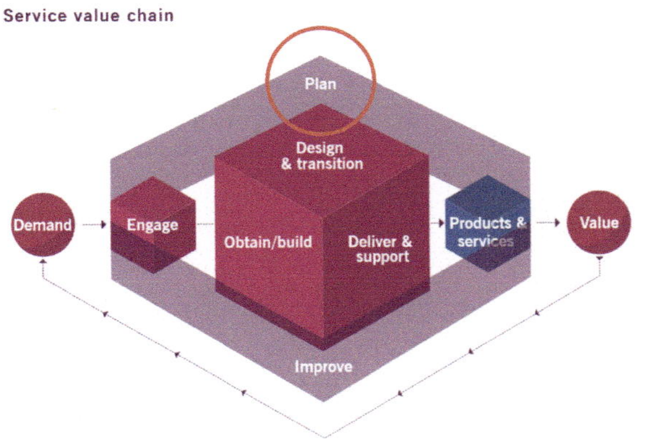

Engage

The purpose is to provide a good understanding of stakeholder needs, transparency, and continual engagement and **good relationships with all stakeholders**

Note: All **incoming and outgoing** interactions with **parties external** to the value chain are performed via engage

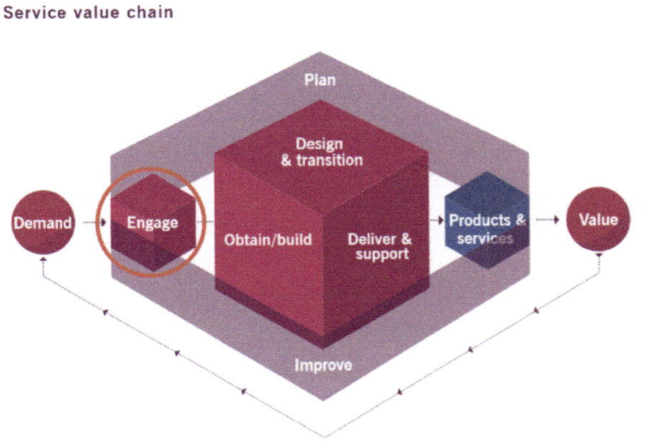

Design and transition

The purpose is to ensure that products and services **continually meet stakeholder expectations** for **quality**, **costs**, and **time to market**

Service value chain

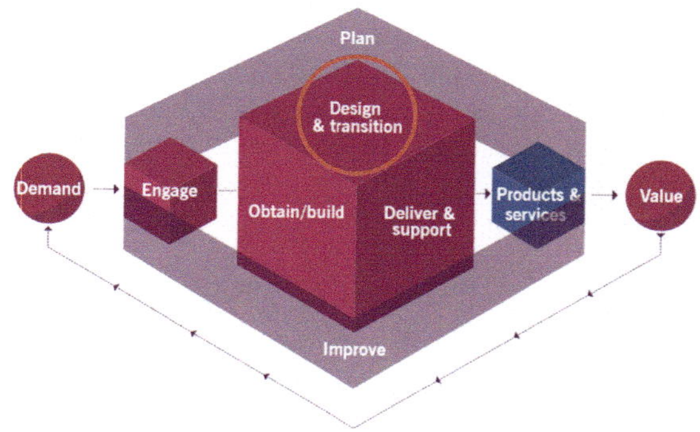

Obtain/build

The purpose is to ensure that service components are **available when** and **where** they are **needed** and **meet agreed specifications**.

Note: All **new resources** are obtained through obtain/build.

Service value chain

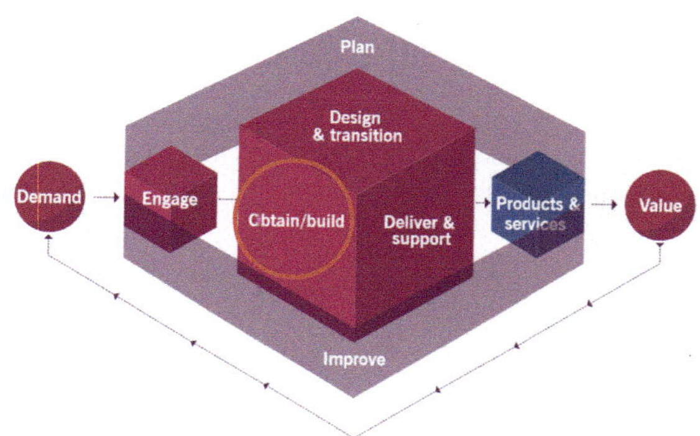

Deliver and support

The purpose is to ensure that services are **delivered** and **supported** according to **agreed specifications** and **stakeholders' expectations**.

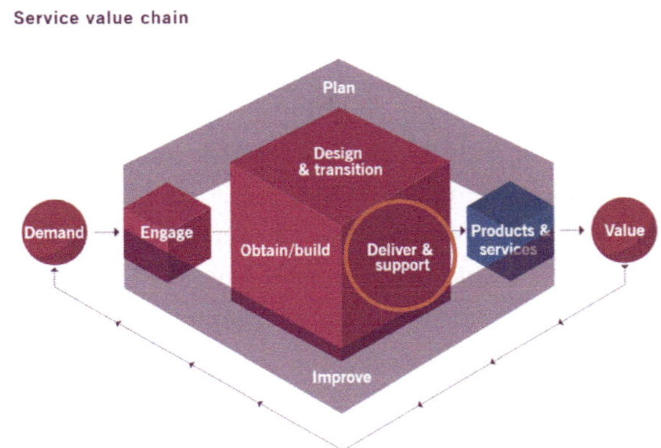

Improve

The purpose of this value chain activity is to **ensure continual improvement** of products, services and practices **across all value chain activities** and the **four dimensions of service management**

Note: Improvements at **all levels** are initiated and managed via improve

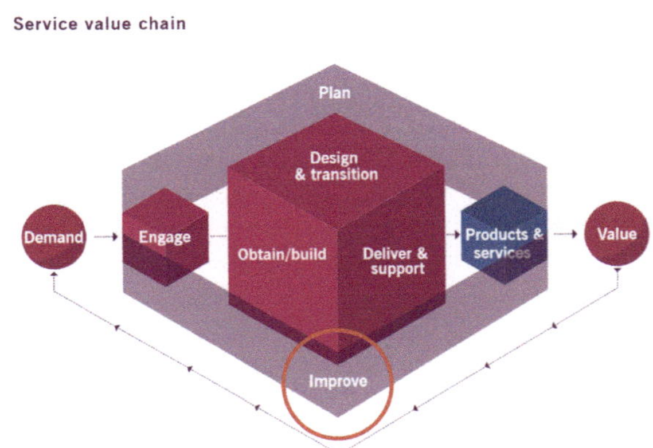

Agile ITSM

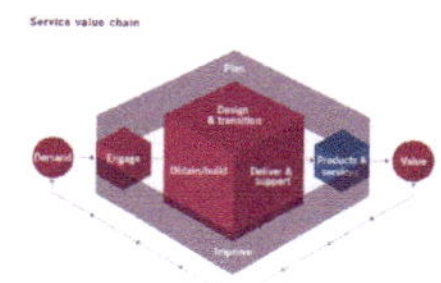

For an organization to be successful, it must be able to **adapt** to **changing circumstances** while **remaining functional** and **effective**. Many principles of Agile development can and should be applied to service management.

Agile software development usually includes:

- **Continually evolving** requirements, collected through feedback analysis and direct observation
- Breaking development work into **small increments and iterations**
- Establishing *product-based cross-functional teams*
- **Visually presented** (Kanban) and **regularly discussed** (daily stand-ups) work progress
- Presenting a working (or the minimum **viable) software** to the stakeholders at **end of each iteration**.

If applied successfully, **agile** software development **enables fast responses** to the **evolving needs** of **service consumers**.

Agile ITSM

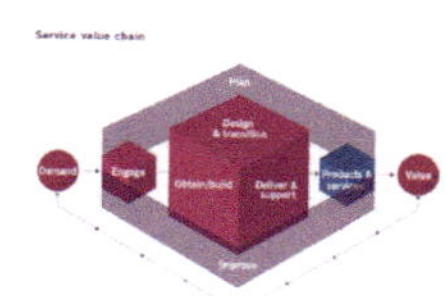

However, in many organizations, agile software development has not provided the expected benefits, often due to **lack of agility in the other phases of the service lifecycle**.

Overall performance of the value chain is **defined by** that of the **slowest part**.

A holistic approach to the service value chain should be adopted to make sure that the service provider is **agile throughout the service lifecycle**.

If adopted as a key principle, agility can **enable** an organization to survive and prosper in a constantly changing environment. **Applied** in a **fragmented** way, agile methods can become a **costly** and wasteful complication.

Summary

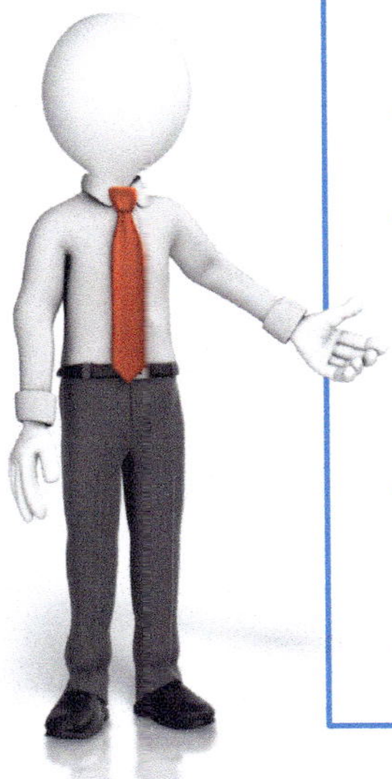

We have just talked about:

- ✓ This section has covered the service value chain, an operating model which outlines the key activities required to respond to demand and facilitate value realization through the creation and management of products and services.
- ✓ The six key activities are:
  - *- Plan*
  - *- Improve*
  - *- Engage*
  - *- Design and transition*
  - *- Obtain/build*
  - *- Deliver and support*
- ✓ The service value streams that are specific combinations of activities and practices, and each one is designed for a particular scenario. Once designed, value streams should be subject to continual improvement.
- ✓ The need for, and positive influence possible, by integrating agile ways of working with ITSM.

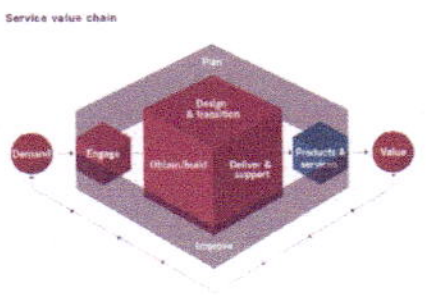

Q: Which value chain activity creates service components?

A. Improve

B. Engage

C. Obtain/build

D. Deliver and support

Q: Which statement about the value chain activities is CORRECT?

A. Every practice belongs to a specific value chain activity

B. A specific combination of value chain activities and practices forms a service relationship

C. Service value chain activities form a single flow that enables value creation

D. Each value chain activity contributes to the value chain by transforming specific inputs into outputs

Q: Which value chain activity includes negotiation of contracts and agreements with suppliers and partners?

A. Engage

B. Design and transition

C. Obtain/build

D. Deliver and support

7 ITIL GUIDING PRINCIPLES

- Understand how the ITIL guiding principles can help an organization adopt and adapt service management

What is an ITIL guiding principle?

It is a recommendation **that guides** an organization **in all circumstances**, regardless of changes in its goals, strategies, type of work, or management structure. A guiding principle is **universal** and **enduring**. But **not prescriptive or mandatory.**

It **embodies the core messages** of ITIL and of service management in general, **supporting** successful **actions** and good **decisions** of **all types** and at **all levels**.

It helps organizations to **adopt** a service management approach and **adapt** ITIL guidance to their **own specific needs** and **circumstances** and encourage and support organizations in **continual improvement at all levels.**

Many of these principles **are reflected** in many **other** frameworks, methods, standards, philosophies and/or bodies of knowledge.

Principle interaction and relevance

ITIL guiding principles, **interact with** and **depend upon each other**

Organizations therefore should not use just one or two of the principles but should ***consider the relevance of each of them and how they apply together***

Not all principles will be critical in every situation, but they should all be reviewed on each occasion to determine how appropriate they are.

Similarly, making use of appropriate **feedback** is key to **collaboration**, and focusing on what will **truly be valuable to the customer** makes it easier to keep things simple and practical.

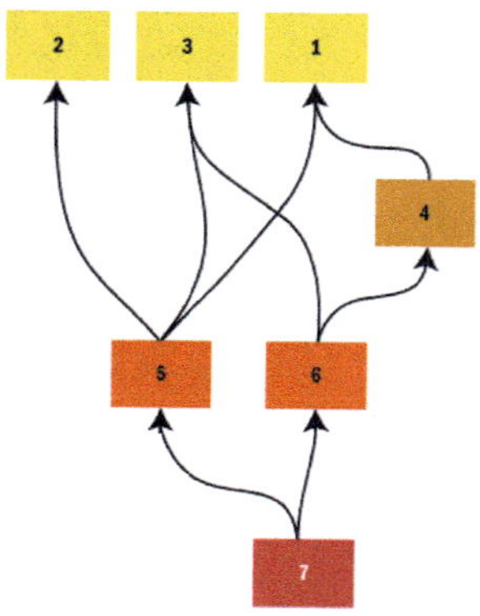

The 7 ITIL guiding principles

Focus on value	**Everything that the organization does needs to map, directly or indirectly, to value for the stakeholders.**
Start where you are	Do not start from scratch and build something new without considering what is already available to be leveraged. The current state should be investigated and observed directly to make sure it is fully understood.
Progress iteratively with feedback	Do not attempt to do everything at once. Even huge initiatives must be accomplished iteratively.
Collaborate and promote visibility	Working together across boundaries produces results that have greater buy in, more relevance to objectives and better likelihood of long-term success. Achieving objectives requires information, understanding and trust.
Think and work holistically	Results are delivered to internal and external customers through the effective and efficient management and dynamic integration of information, technology, organization, people, practices, partners and agreements, which should all be coordinated to provide a defined value.
Keep it simple and practical	If a process, service, action or metric provides no value, or produces no useful outcome, eliminate it. Always use outcome-based thinking to produce practical solutions that deliver results.
Optimize and automate	Eliminate anything that is truly wasteful and use technology to achieve whatever it is capable of. Human intervention should only happen where it really contributes value.

The Agile methodology

A methodology that focuses on **delivery** and **evolution** of requirements from **small teams**. It is a **timeboxed**, flexible and adaptive approach focusing on **rapid** response to change.

Teams are often **self organized** and there is a high degree of **collaboration** between customers, users and development teams at every opportunity.

Possible challenges with Agile are:

- Basic **organizational** capabilities need to be established.
- **Focus on features and fixes**, rather than improving the service.
- Higher **costs over time**, less accurate time estimations and a lower quality of service delivery.
- It can result in **over-engineered** projects that have been built iteratively with no consideration for the overall service.

Agile Method

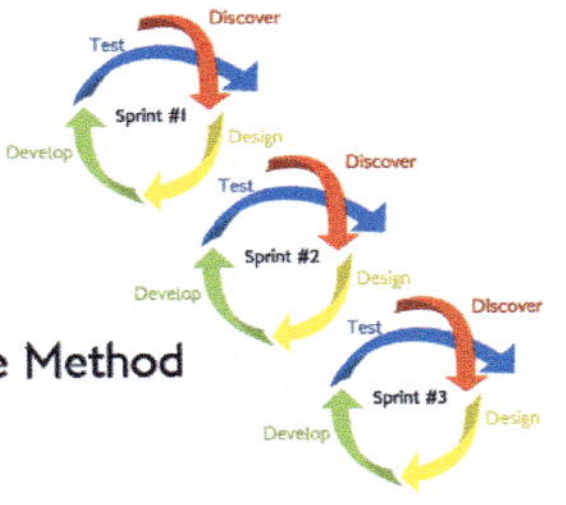

ITIL and Agile in cooperation

ITIL can complement Agile teams to enable more effective, faster and stable deployment within the **context of the wider service concept** and reduce ongoing cost of the service.

Together they they can lead to better **co-ordination** between agile projects and other **areas** of the service/business that might **not work with Agile methods**.

Agile shares many common themes with ITIL, and together they can support each other to produce highly **effective practices** including, but not limited to:

- **Continual improvement**
- **Problem management**
- **Change control**
- **Release management**
- **Deployment management.**

ITIL and DevOps in cooperation

DevOps arose from the increasing success of Agile software projects that led organizations to **release more frequently**

DevOps centers on the process of delivering software to live environments, with a focus on **unifying technical operations and delivery**.

Anyone with the responsibility for **change control**, **release management** and **service operations** should be a **part of the Agile team** to ensure that the team work effectively together.

All coming together

Agile methods focus on working in timeboxed iterations, which can carry a risk of introducing instability in ongoing services. As Agile focuses on automation and speed of delivery it may be beneficial for teams to also use DevOps to help ITIL and Agile practices work well together.

The Agile roles within teams can be multi-purpose and aligned to ITSM roles, for example:

- Product managers/owners can perform the role of the service owner
- Scrum masters can perform the role of the change manager
- Scrum masters are already running retrospectives and ensuring lessons are learned, which can form a part of the wider continual improvement practice.

ITIL and Agile can be great allies. An Agile team focusing on customer needs and satisfaction through the lens of the overall service will deliver greater value in a shorter amount of time.

Sets of principles compared

Agile Manifesto	ITIL guiding principles
Individuals and interactions *over* processes and tools	• **Keep it simple and practical** • **Start where you are**
Working software *over* comprehensive documentation	• **Focus on value** • **Think and work holistically**
Customer collaboration *over* contract negotiation	• **Focus on value** • **Collaborate and promote visibility**
Responding to change *over* following a plan	• **Progress iteratively with feedback** • **Keep it simple and practical**

Focus on value

Everything the organization does, should link back, directly or indirectly, to **value** for itself, its customers and other stakeholders

Value may come in various forms, such as revenue, customer loyalty, lower cost or growth opportunities

Who is the service consumer?

When focusing on value, the **first step** is to know **who is being served**. In each situation the service provider must, therefore, determine **who the service consumer** is and **who the key stakeholders** are, in doing this the service provider should consider **who will receive value** from what is being delivered or improved.

Note: Remember to address various stakeholder groups, ***not only customers****.*

Consumer perspective

When identified the consumer, the service provider needs to further know:

- **why** the consumer uses the services?
- **what** the services help them to do?
- **how** the services help them meet their goals?
- the role of **cost/financial** consequences for the service consumer?
- the role of **risks** for the service consumer?

Value is defined by the services consumer's **own needs,** achieved through the support of **intended outcomes** and **optimization** of the **costs** and **risks**. It also **changes over time** and **in different circumstances**.

The customer or user experience

An **important element** of value is the **experience** consumers have when they interact with the service and the service provider. This is often further expanded into customer experience (**CX**) or user experience (**UX**).

CX (or UX) can be defined as the entirety of the interactions a customer (or user) has with an organization and its products. This experience can **determine how the customer feels** about the organization and its **products** and **services**. This is **both objective and subjective**.

For example:

When a customer orders a product and receives what they ordered at the promised price and in the promised delivery time, the success of this aspect of their experience is objectively measurable. On the other hand, if they don't like the style or layout of the website they are ordering from, this is subjective. Another customer might really enjoy the design.

Applying the principle

Know how service **consumers use** each service

- *Collect feedback on value on an ongoing basis, not just at the beginning of the service relationship*

Encourage a **focus on value among all staff**

- *Teach staff to be aware of who their customers are and to understand CX (and UX)*

Focus on value during normal **operational** activity as well as during **improvement** initiatives

- *The organization as a whole contributes to the value that the customer perceives, and so everybody within the organization must maximize the value they create*

Include **focus on value in every step** of any improvement initiative

- *Everybody involved in improvement needs to understand what outcomes the initiative is trying to facilitate, how its value will be measured, and how they should be contributing to the co-creation of value*

Start where you are

In the process of eliminating old, unsuccessful methods or services and creating something better, there can be great **temptation to remove** what has been done in the past and build something completely new

This is **rarely necessary**, or a wise decision. On the contrary, it can be **extremely wasteful**, not only in terms of time, but also in terms of the loss of existing services, processes, people, and tools that could have significant value in the improvement effort

Do not start over without first considering what is already available to be leveraged

Assess where you are

Services and methods already in place should be measured and/or observed directly to properly understand their current state and what can be reused from them.

Ensure decisions are based on **accurate** information. Be aware that there is often **discrepancy** between reports and reality.

Getting **data from the source** helps to avoid assumptions.

To ensure this, remember:

- To observe activities directly
- That there are no stupid questions
- That different roles need to be part of the observations, and get firsthand information

The role of measurement

The **use of measurement** is important to the principle "start where you are" Use it to support the analysis of what has been observed rather than to replace it, as over-reliance on data analytics can introduce biases and risks

Organizations should consider a **variety of techniques to get knowledge** of their working environments. Some things can **only** be understood **through measuring** their effect, but **direct personal** observation should always be preferred.

Note: The act of **measuring can** sometimes **affect the results**, making them inaccurate. **People are very creative** in finding ways **to meet the metrics** they are measured against. Therefore, metrics need to be meaningful and directly relate to the desired outcome.

> "When a measure becomes a target, it ceases to be a good measure"
> Goodhart's law

Applying the principle

Having a proper understanding of the current state of services and methods is important to selecting which elements to reuse, alter or build upon

Consider these advices:

- Look at **what exists** as objectively as possible, using the desired outcome **as a starting point**
- When examples of **successful** practices or services are **found**, **determine** if and **how** these can be **replicated** to achieve the **desired state**
- Apply your **risk management skills**
- Recognize that **sometimes nothing** from the current state **can be re-used**.

Progress iteratively with feedback

Resist the temptation to do everything at once. By **organizing work into smaller**, manageable sections that can be executed and completed in a **timely manner**, the focus on each effort will be sharper and easier to maintain **sequential or simultaneous**

Each individual **iteration** should be both manageable and managed, ensuring that tangible results are returned in a timely manner and **built upon** to create **further improvement**

The overall program, as well as its component iterations, must be **continually re-evaluated** and potentially revised to reflect any changes and ensure that focus on value has not been lost

This re-evaluation should make use of a **wide range of feedback channels and methods** to ensure that the status of the initiative and its progress are properly understood.

The role of feedback

No improvement iteration occurs in a vacuum. Whether working on improving a service, group of services, practice, process, technical environment or other service management element. **Circumstances** can **change** and new priorities arise, and the need for the iteration may change. Seek to use **feedback before**, **throughout** and **after each iteration**

A **feedback loop** is a situation where part of the output of an activity is used for new input. Feedback should **actively** be collected and processed **along the value chain** to understand:

- end user and customer **perception** of the value created
- the **efficiency** and **effectiveness** of value chain activities
- the effectiveness of service **governance** as well as **management** controls
- the **interfaces** between the organization and its partner and supplier network
- the **demand** for products and services

Iteration and feedback together

Working in a **time-boxed**, **iterative** manner with **feedback loops** embedded into the process allows for:

- **Greater flexibility**
- **Faster** responses to customer and business needs
- The **ability to discover** and respond to failure earlier
- An overall improvement in **quality**.

*Appropriate feedback loops between the participants of an activity gives them a better understanding of where their work comes from, where their outputs go and how their actions and outputs affect the outcomes, which in turn enables them to **make better decisions.***

Applying the principle

To apply this principle successfully, consider these advices:

- **Comprehend the whole, but do something**
  Sometimes the greatest enemy to progressing iteratively is the desire to understand and account for everything. Understanding the big picture is important, but so is progress
- **The ecosystem is constantly changing, so feedback is essential**
  Change is happening constantly, so it is important to seek and use feedback at all times and at all levels
- **Fast does not mean incomplete**
  Just because an iteration is small enough to be done quickly does not mean that it should not include all the elements necessary for success. Create a minimal viable product, or a version which allows the maximum amount of learning with the least effort

Collaborate and promote visibility

Involve the right people in the correct roles, efforts benefit from better buy-in, more relevance (because better information is available for decision-making) and increased likelihood of long-term success

- **Conclusion** is generally a better policy than exclusion.
- **Silos** can occur through the behavior of individuals and teams, but also through structural causes
- **Applying** the guiding principle of think and work holistically can help organizations to break down barriers between silos of work
- **Recognition** of the need for genuine collaboration

Without effective collaboration, neither DevOps, *Agile, Lean nor any other ITSM framework or method will work.*

Transparency builds trust

Working together in a way that leads to real accomplishment requires **information**, **understanding** and **trust**.

Work and its results should be made visible, hidden agendas should be avoided and **information should be shared** to the greatest degree possible.

The more **people are aware** of what is happening and why, the more they will be willing to help. When activity's occurs behind closed doors, assumptions and rumors can prevail.

Resistance to change will often rise as staff members are unaware and therefore speculate about what is changing and how it might impact them.

Whom to collaborate with

Working together in a way that leads to real accomplishment requires **information**, **understanding** and **trust**.

Work and its results should be made visible, hidden agendas should be avoided and **information should be shared** to the greatest degree possible.

The more **people are aware** of what is happening and why, the more they will be willing to help. When activity's occurs behind closed doors, assumptions and rumors can prevail.

Resistance to change will often rise as staff members are unaware and therefore speculate about what is changing and how it might impact them.

Other stakeholders to involve:

Developers working with other internal teams to ensure that what is being developed can be operated efficiently and effectively. Collaborate with technical and non-technical operational teams.

Suppliers collaborating with the organization to define its requirements and brainstorm solutions to customer problems - Internal and external suppliers collaborating with each other.

Relationship managers collaborating with service consumers to achieve a comprehensive understanding of service consumer needs and priorities.

Customers collaborating with each other to create a shared understanding of their business issues.

Communication for improvement

For each stakeholder group, it's important to define the **most effective methods to use**

For example,
the contribution to improvement from customers of a public cloud service may be through a survey or checklist of options for different functionality. For an internal customer group, the contribution to improvement may come from feedback solicited via a workshop or a collaboration tool on the organization's intranet.

Some contributors may need to be involved at a very **detailed level**, while **others** can simply be involved as **reviewers or approvers**

Depending on the service and the relationship between the service provider and the service consumer, the expectations about the level and type of collaboration can **vary significantly**

Increasing urgency through visibility

When stakeholders **have poor visibility** of the workload and progression of work, the **impression is that the work is not a priority.** Equally, when staff members attempt to prioritize improvement over business-as-usual (BAU), **it may seem to be a low priority** without support by the management.

Poor visibility of work leads to poor decision-making. To avoid this organizations need to:

- understand the flow of work in progress
- identify bottlenecks, as well as excess capacity
- uncover waste

*It is important to involve and address the needs of stakeholders at all levels. How it relates to the stated **vision, mission, goals and objectives** of the organization. Determining the **type**, **method** and **frequency** of such **messaging** is one of the central activities in communication*

Applying the principle

Collaboration does not mean consensus

It is not necessary. Some organizations are so concerned with getting consensus that they try to make everyone happy and end up either doing nothing or producing something that does not properly suit anyone's needs.

Communicate in a way the audience can understand

Selecting the right method and message for each audience is critical for success.

Decisions can only be made on visible data

Making decisions in the absence of data is risky. There may be a cost to collecting data, and the organization must balance that cost against the benefit and intended usage of the data.

Think and work holistically

No service, practice, process, department or supplier stands alone
The outputs that the organization delivers to itself, customers and other stakeholders will suffer unless it works in an integrated way to handle its activities as whole, rather than as separate parts. All the organization's activities should be focused on the delivery of value

Services are delivered to internal and external service consumers through the coordination and integration of **the four dimensions of service management.** Taking a holistic approach to service management includes establishing an understanding of how all the parts of an organization work together in an integrated way. It requires end-to-end visibility of how demand is captured and translated into outcomes

In a complex system, the alteration of **one element can impact others** and, where possible, these impacts need to be identified, analyzed and planned for

Applying the principle

Recognize the complexity of the systems. Different levels of complexity require different heuristics for decision-making. Applying methods and rules designed for a simple system can be ineffective or harmful in a complex system

Collaboration is key to thinking and working holistically If the right mechanisms are put in place for all relevant stakeholders to collaborate in a timely manner

Look for patterns in the needs of and interactions between system elements to identify what is essential for success. With this information, needs can be anticipated, standards can be set and a holistic view point can be achieved

Automation can support end-to-end visibility for the organization and provide an efficient means of integrated management

Keep it simple and practical

Always use the **minimum number of steps** needed to accomplish an objective. Outcome based thinking should be used to produce **practical solutions** that deliver **valuable outcomes**

If a process, service, action or metric provides **no value** or produces no useful outcome, then **eliminate it**

Although this principle may seem obvious, it is **frequently ignored**

When creating a process or a service, designers need to **think about exceptions**, but they cannot cover them all. Instead, rules should be designed that can be used to **handle exceptions generally**

Conflicting objectives

When designing, managing, or operating practices, be mindful of **conflicting objectives**

For example,

The management of an organization may want to collect a large amount of data to make decisions, whereas the people who must do the record-keeping may want a simpler process that does not require as much data entry

Through the application of this, and the other guiding principles, the organization should agree on a **balance between its competing objectives**

For example,

Services should only generate data that will truly provide value to the decision-making process, and record keeping should be simplified and automated where possible to maximize value and reduce nonvalue-adding work

Applying the principle

Advice to keep it simple and practical:

- **Ensure value** - Every activity should contribute to the creation of value
- **Simplicity is the ultimate sophistication**
- **Do fewer things, but do them better** - Minimizing activities to only include those with value for one or more stakeholders more focus on the quality of those actions
- **Respect the time of the people involved** - A process that is too complicated and bureaucratic is a poor use of the time of the people involved
- **Easier to understand, more likely to adopt** - To embed a practice, easy to follow
- **Simplicity is the best route to achieving quick wins** - Whether in a project, or when improving daily operations activities, quick wins allow organizations to demonstrate progress and manage stakeholder expectations. Working in an iterative way with feedback will quickly deliver incremental value at regular intervals

Optimize and automate

Organizations must **maximize** the **value** of the **work** carried out by its human and technical resources.

The **four dimensions model** provides a holistic view of the various constraints, resource types, and other areas that should be considered when designing, managing, or operating an organization.

Technology can help organizations to scale up, However, **technology should not always be relied upon** without the capability of **human intervention.**

Before it can be **effectively automated**, activities should be **optimized**. To optimize means to make something as effective and useful as it needs to be, within a set of constraints.

The road to optimization

Practices of continual improvement and **measurement** and **reporting**, are **essential** to this effort. Improve and optimize performance may draw upon **guidance from** ITIL, Lean, DevOps, Kanban and other sources.

Regardless of the specific techniques, the **path** to optimization follows these **high-level steps**:

1. Understanding and agreeing the context and the overall vision and objectives of the organization
2. Assessing the current state to understand where it can be improved and which likely to produce the biggest positive impact
3. Agreeing what the future state is and priorities should be, focusing on simplification and value. This typically also includes standardization of practices and services, which will make it easier to automate or optimize further at a later point
4. Ensuring the optimization has the appropriate level of stakeholder engagement and commitment
5. Executing the improvements in an iterative way, using metrics and other feedback to check progress
6. Continually monitoring the impact of optimization to identify opportunities to improve methods of working

Using automation

Automation typically refers to the **use of technology** to perform a step or series of steps correctly and consistently with limited or no human intervention

For example, in organizations adopting **continuous deployment**, through to **live**, and often **automatic testing** occurring in each environment

In its simplest form, however, automation could also mean the standardization and stream-lining of **manual tasks**. Efficiency can be greatly increased by **reducing** the need for **human involvement** to stop and evaluate each part of a process

Opportunities for automation can be found across the entire organization to automate **standard** and **repeating tasks** can help save the organization **costs**, reduce human **error**, and improve employee **experience**

Applying the principle

Simplify and/or optimize before automating

- Take time to map out the standard and repeating processes as best as possible, and streamline where possible (optimize). From there you can start to automate

Define your metrics

- Use the same metrics to define the baseline and to measure the achievements and make sure that the metrics are outcome based and focused on value

Use the other guiding principles when applying this one

- **Progress iteratively with feedback** Iterative optimization and automation make progress visible and increase stakeholder buy-in
- **Keep it simple and practical** Things can be simple, but not optimized, so use the principles together
- **Focusing on value** Selecting what to optimize and automate and how should be based on best value
- **Start where you are** Make use of what is already there, but untapped or underutilized, to implement opportunities for optimization and automation quickly and economically

Summary

We have just talked about:

- ✓ This section has covered the ITIL guiding principles and how they can be used in all circumstances, regardless of changes in goals, strategies, type of work, or management structure. A guiding principle is universal and enduring.
- ✓ The ITIL guiding principles are:
  - *Focus on value*
  - *Start where you are*
  - *Progress iteratively with feedback*
  - *Collaborate and promote visibility*
  - *Think and work holistically*
  - *Keep it simple and practical*
  - *Optimize and automate*
- ✓ All principles integrate and can and should be used in conjunction with other available frameworks or best practices like e.g. Agile, DevOps, COBIT or Lean.

Q: Which describes the nature of the guiding principles?

A. A guiding principle can guide an organization in all circumstances

B. Each guiding principle mandates specific actions and decisions

C. An organization will select one of the principles to adopt

D. Guiding principles describe the processes that all organizations must adopt

Q: Which is a key consideration for the guiding principle 'keep it simple and practical'?

A. Try to create a solution for every exception

B. Understand how each element contributes to value creation

C. Ignore conflicting objectives of different stakeholders

D. Start with a complex solution, then simplify

Q: Which guiding principle recommends organizing work into smaller, manageable sections that can be executed and completed in a timely manner?

A. Focus on value

B. Start where you are

C. Progress iteratively with feedback

D. Collaborate and promote visibility

Q: What is the first step of the guiding principle 'focus on value'?

A. Identify the outcomes that the service facilitates

B. Identify all suppliers and partners that are involved in the service

C. Determine who the service consumer is in each situation

D. Determine the cost of provisioning the service

Course schedule

- ✓ Day 1: ITIL 4 Foundation
  - Key concepts, 4 dimensions of SM, ITIL Service Value System, the service value chain & 7 guiding principles
- ✓ Day 2: Create, Deliver and Support
  - How to use Value streams with support of ITIL practices
- ✓ Day 3: Drive Stakeholder Value
  - Customer journeys, relationships and service offerings
  - How to act together to ensure continual value co-creation
- ✓ Day 4: High Velocity IT
  - The Digital enterprise and how to use guiding principles
- ✓ Day 5: Deliver, Plan and Improve and Exam
  - Direction and Planning, GRC, Communication & OCM
  - Work with sample exam and examination

ITIL 4:
CREATE, DELIVER AND SUPPORT

Copyright © Van Haren Publishing and AXELOS Limited This material is based on AXELOS ITIL® material.
Material is reproduced under license from AXELOS. All rights reserved

Key learning requirements

CREATE, DELIVER & SUPPORT

- Understand how to plan and build a service value stream to create, deliver, and support services
- Know how relevant ITIL practices contribute to the creation, delivery and support across the SVS and value streams
- Know how to create, deliver and support services

Introduction

Create Deliver and Support covers the integration of a number of proven areas that often have not been built, run and integrated as seamlessly as needed to fully deliver optimum value. The core proposition of ITIL 4 is that this work is all part of a single value chain. This book includes core content describing how different types of work (value streams) are built, tested and delivered 'end-to-end' from beginning to end and with continual iterations and feedback loops

Key terms defined and important concepts of service management introduced here include:

- organizational structures and integrated/collaborative teams
- team capabilities, roles, competencies, culture and differences
- a customer-oriented mindset and the culture of continual improvement
- how to use value streams to design, develop, transition and support services
- how to co-ordinate, prioritize and structure work to create, deliver and support services.

Organizational structure

Definition: Organization

A person or a group of people that has its own functions with responsibilities, authorities, and relationships to achieve its objectives.

There are various types of organizational structures. Some high-level examples are:

- **Functional** – often based on functional areas with formal lines & roles
- **Divisional** – often based on markets, products, geography, etc. with each having profit & loss responsibility
- **Matrix** – a grid of relationships, often dual reporting lines, can provide speed and agility
- **Flat** – very little hierarchy, often fast decision making, can be challenging to maintain as organization grows

Organizational agility and resilience

Organizational agility is the ability of an organization to move and adapt quickly, flexibly and decisively to support internal changes.

Organizational resilience is the ability of an organization to anticipate, prepare for, respond to and adapt to both incremental changes and sudden disruptions from an external perspective. It requires a common understanding of priorities and objectives.

Successful organizations must achieve agility and resilience to support internal changes, and withstand or even thrive in changing external circumstances. They must also be part of larger ecosystems, delivering, coordinating and consuming products and services.

The SVS provides the means to achieve organizational agility and resilience and to facilitate the adoption of a strong unified direction, focused on value and understood by all.

Adopting organizational structure

The ITIL 4 guiding principles are a useful reference point when planning to change and improve the organisational structure. The following considerations may be useful:

- Focus on value: What is the key driver for changing the way of working?
- Progress iteratively with feedback: Ensure that the work is broken down into manageable "bite size"
- Start where you are: Ensure that the cultural aspects of the organisation are considered as part of this design. Use value stream mapping and RACI matrix to understand the current roles and responsibilities
- Collaborate and promote visibility: Ensure that all stakeholders are engaged throughout the change process and enable transparency
- Think and work holistically: Collaboration with all appropriate leaders/managers will ensure any potential risks are understood and managed as appropriate
- Keep it simple and practical: Reduce the complexity of the organisation as much as possible to enable the flow of work and information to be uninhibited
- Optimize and automate: Design and plan in a way where any tasks can be consolidated or automated

A collaborative culture

Cooperation

- Working with others to achieve your goals, which may be part of a common goal
- May be an effective model for standardized work in work with clear separation of duties
- Everyone needs to understand their own role
- Individuals and teams succeed independently
- Need to understand PESTLE factors for own role
- Less need for trust and transparency
- Needs for effective communication
- Alignment of goals

Collaboration

- Working together to achieve a common/shared goal
- Higher potential for creative and entrepreneurial work in complex environments
- Everyone needs to understand how they contribute to the big picture
- Everyone succeeds or fails together
- Need to understand PESTLE factors for all stakeholders
- Needs for respect, trust and transparency
- Needs multi-channel communication
- Shared and integrated goals

Algorithmic or heuristic work?

Algorithmic work:

Involves a person following a defined process, driven by a set of established instructions, along a consistent pathway until the work is concluded

- Predictable process paths
- Clear inputs, instructions, outputs, etc.
- Reassignments and handovers
- All roles should should be encouraged to identify opportunities for improvement

Heuristic work:

Fundamentally more dependent on human inventiveness. It involves enabling a person to discover or learn something for themselves

- Depends on human understanding & intervention
- Learning and discovery
- Needs flexibility, information, knowledge & experience
- Insights can be recorded for future use, moving some work to algorithmic (removing 'toil')

The service designer needs to understand the nature of the work on which their service and process depends

Design for collaboration

Gain knowledge from collaboration!

In heuristic situations, new insights and solutions will frequently emerge in the course of work getting done

Even where the algorithmic approach applies, companies have found great benefit in making it easy for the workers performing the tasks suggest improvements

Collaboration frameworks should be used which enable the knowledge gained to be captured, refined and reused

Effective design of collaboration and workflow requires each interaction to align well to the needs of the agents involved

Servant Leadership

In order to adapt to new more flexible and responsive ways of working such as Agile and DevOps, many organisations have adopted new approaches, where a leader's role is more 'servant'

Key assumptions in 'Servant Leadership':

- Managers should meet the **organization's needs**, not only those of teams
- Managers are there to **'serve'** and **support** their people and ensure they have the right resources and support to get their jobs done

Servant Leadership can be often seen in flat, matrix, or product-focused organizations; however, this approach is applicable and beneficial in any organizational structure. It inspires people to naturally collaborate and use the leaders support to become more cohesive and productive

Roles and competencies

Roles in IT traditionally have followed technical competencies and been related to clearly defined areas of work in development or operations

In the information age roles tend to change more often and require more flexibility

Even IT and ITSM roles now require soft skills and business capabilities such as:

Ability to manage and motivate people and teams

Relationship management

Negotiation skills

Supplier and contract management

It is important that relationships with all key stakeholders are understood and managed.

Professional ITSM competencies

Examples of skills and knowledge needed to be fully effective in service management:

- **Communication skills** – ability to establish good working relationships with a variety of stakeholders at a number of levels, using different techniques
- **Business and commercial skills** – Service managers are required to use a number of commercial skills to specify, buy, negotiate and manage these relationships
- **Relationship management skills** – an active liaison and synchroniser for information, feedback, demand and progress between parties
- **Leadership skills** – ability to influence, motivate and support people in their work and build a culture
- **Market and organisational knowledge** – being aware of the business and market sector of their organisation including industry factors, competition and cost & price considerations
- **Management and administration** – good management, teambuilding, recruitment, delegation, documentation and logistical skills
- **Developing innovation** – required to identify new ways of working, delivering services and solving problems

Neither generalist or specialist

Increasingly organisations look for people of these types:

- T-shaped - Individuals with expertise in one area and broad, less profound, knowledge in other areas
- Pi (π) shaped - Individuals with expertise in two areas and broad, less profound, knowledge in other areas
- Comb ( ) shaped - Individuals with expertise in more than two areas and broad, less profound, knowledge in other areas

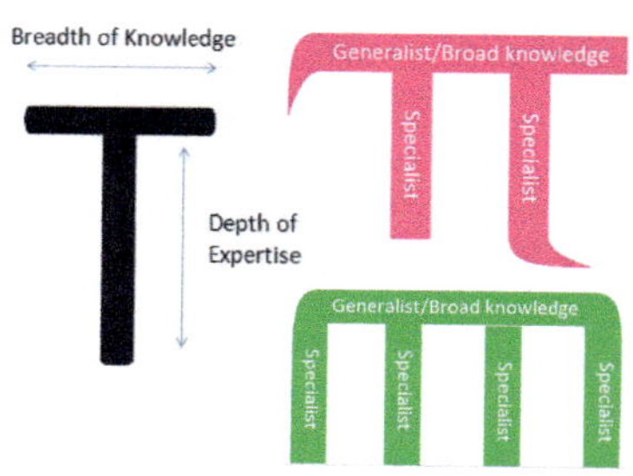

Although a clear focus on one competence creates deeper understanding, it can be dangerous to have just one area of profound expertise

Developing broad competencies

Examples of different ways of gaining and recognizing a wider set of competencies:

- Building job descriptions, that clarify the full non-technical requirements for roles
- Developing recruitment and on-boarding skills
- Recognising non-IT experience in – e.g. team management, procurement, contract management
- Ensuring that job descriptions include 'soft' skills – communications, leadership, innovation
- Performance management, appraisals and reward programmes should reflect and reward the full scope
- Opportunities for training and development for all staff in all areas
- Encouraging CPD (Career Professional Development) initiatives that recognise and develop all areas
- Role-based and competency-based models, based on job descriptions with career paths
- Competency based models that are more focussed on generic skills
- Hybrid role-based and competency-based models by combining job related and talent related aspects

There is no single path to achieving a full set of competency across all areas!

Developing a team culture

What is a 'team culture'?

- A set of **values** that are shared by a group of people, including **expectations about** how people should **behave**, **ideas**, **beliefs**, and **practices**

Service providers focusing on value creation display these common characteristics:

- Value, quality and operational excellence focus
- Client, customer and consumer orientation
- Investment in people and communication/collaborative tools
- Strong team composition within a structured organisation
- Continuous alignment with the vision, mission and strategic objectives

Often it is the unwritten – 'how we do things here' – or 'how we've always done things here'

Cultural fit

What is cultural fit and why is it important?

- Cultural fit is the ability for an employee or a team to comfortably work in an environment that is in line with their own beliefs, values, and needs

Note: When hiring for cultural fit, it is important to be aware of bias. It is human nature to gravitate towards like-minded individuals with a similar personality or beliefs

How to nurture good team culture

It is possible to grow and evolve a teams culture over time. ITIL Guiding Principles and the continual improvement model can be very useful tools for implementing cultural change

Some simple guidelines on how to develop and nurture good team culture:

- Incorporating the vision into the team culture
- Meeting regularly
- Creating leaders, more than managers
- Encouraging informal teams
- Cross-training employees
- Integrating socially
- Providing feedback
- Promoting a culture of learning

Continual improvement culture

Continual improvement should not be thought of as a practice or process, as a workflow or just as some key artefacts or tools – it should be embedded in all of these things, as the ***'way we do things here'*** – as the culture of the organisation

Some key elements of a continual improvement culture are:

- Transparency
- Management by example
- Building trust
- Active encouragement of positive behaviours
- Clear continual improvements expectations
- Marketing and celebration of success

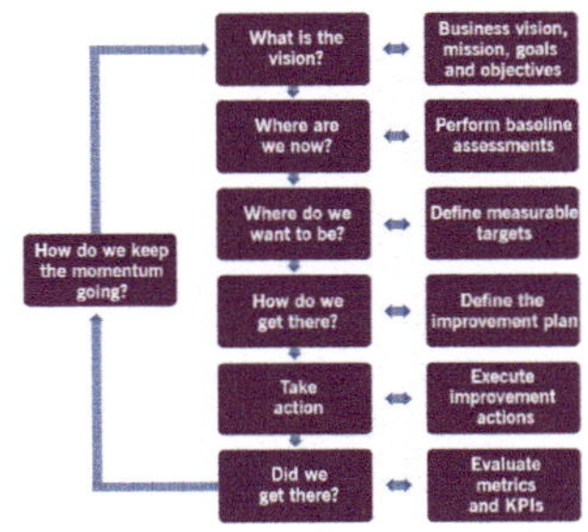

Creating this environment within an existing culture is a challenge and requires bravery, commitment and persistence and needs ongoing, visible support from senior management

Service empathy and mindset

It is critical to the success of service relationship that all involved in the service provision and consumption act responsibly, considering the interests of others and focusing on the agreed service outcomes

Service empathy

- The ability to recognize, understand, predict, and project interests, needs, intentions and experience of another party, in order to establish, maintain and improve service relationship

Service empathy is an element of service mindset, but not the only one. Service mindset also includes shared principles that drive organization's behaviour and define organization's attitude towards the service relationship and other parties involved

Service mindset

- An important component of organizational culture, which defines organization's behaviour in service relationships. Service mindset includes shared values and guiding principles adopted and followed by the organization

A customer-oriented mindset

From a service provider perspective, service mindset is also known as customer orientation
A customer-oriented organization places **customer satisfaction at the core** of each of its business decisions

It is defined as an approach to sales and customer-relations in which staff focus on **helping customers** to meet their **long-term needs** and wants. Here, management and employees **align** their individual and team objectives **around satisfying** and retaining **customers**

Strive to:

- Understand that every customer is **unique** and have **individual needs** and wishes
- Create a **clear link** between your **service** or product and these **customer needs**
- **Focus on value** and on the correct 'package' of customer needs
- Care about **customer experience** and continually improve it

It is important that relationships with <u>all</u> key stakeholders are understood and managed.

Customer experience

Steps to help an organisation become customer oriented:

- Create a value proposition that sells the organisation and services
- Map the customer/user experience journey and all its touchpoints
- Recruit customer-friendly people
- Treat employees well
- Train and coach people and teams
- Lead by example
- Listen to the Voice Of the Customer (VOC)
- Use a broad set of inputs and channels for feedback
- Empower staff
- Avoid silo mentality and encourage close cooperation

It is important that relationships with <u>all</u> key stakeholders are understood and managed.

Manage employee satisfaction

The true potential of an organization can only be realized when the productivity of individuals and teams are aligned, and their activities integrated to achieve the goals of the organization

Understand and manage employee satisfaction

- Morale and engagement influence productivity, customer satisfaction, employee retention...

Need to measure many aspects of employee satisfaction

- Leadership, culture, organization climate, job activities
- Baseline satisfaction levels
- Identify actions to improve commitment and trust

Don't just measure, improve, and be seen to improve

- Assign resources to act on the measurement
- Results and actions must be demonstrated back to employees

Key elements in feedback:

- *Confidentiality*
- *Support & understanding*
- *Call to action*

Feedback types

There are many ways of measuring employee satisfaction. Some typical feedback types are:

- Employee surveys
- Regular meetings
- Unstructured meetings
- Reviewing sickness and attrition data
- Staff driven metrics

One challenge for organizations conducting surveys is **presenting the results** and **corresponding actions** based on the analysis of the results. Feedback to users need to be timely, as well as the actions taken in response to the results

It is important to remember that the top **four motivation factors** for people at work are **achievement**, **recognition**, **responsibility** and **interesting work**

Value of positive communication

The ability to communicate effectively is a key business skill, which is also fundamental to success in service management

Good human communication is about being effective, often through establishing positive relationships, which include trust, empathy, proximity and shared goals that avoid problems and stress. This forms the basis for the success of the services that are delivered

Good communication starts with listening. Try to understand, more than being understood

Good communication is good for business

Communication principles

The fundamental principles required for good communication can be summarised as:

- **Communication is a two-way process** – successful communication is an exchange of information and ideas between two or more parties
- **We are all communicating, all the time** – people convey messages about their mood, attitude and emotional state through the use of language, tone of voice, body language, dress and manners
- **Timing and frequency matters** – successful communications needs to take into account when is the best time to make contact and also when is not
- **There is no single method of communicating that works for everyone** – its important to recognise and make use of different preferences and methods used by those with whom we interact
- **The message is in the medium** – choose an appropriate way that is appropriate for the importance of the message that is being communicated. A minor point may be communicated via messaging or email. Big issues or questions require direct discussion and should not be carried out via email

Understanding, recognising and acting upon these principles is vital in order to build positive relationships with colleagues, customers and all stakeholders

Team collaboration & integration

Cooperation and collaboration are vital for effective and valuable teamwork and service relationships

It is impossible to force collaboration, as it is based on shared goals and a high level of trust

Sometimes it is more realistic to establish effective cooperation within and between teams using aligned, transparent, and integrated goals and metrics

Shared principles can be a foundation for effective teamwork and a starting point for its improvement

Workforce planning & management

The workforce is arguably the most important asset of most organisations. It is essential to identify the required roles, associated knowledge, skills and attitudes to:

- Keep the organisation running
- Exploit emerging technologies
- Provide leadership and organisational change capabilities
- Position the organisation for future success and growth

Fundamentally, workforce planning and talent management are a set of specific workforce strategies for recruiting, retaining, developing, and managing employees

The purpose of the **workforce and talent management practice** is to enable organization, leaders, and managers to focus on creating an effective and actionable people strategy so that the organization can achieve its mission, goals, and strategic objectives

Applied to the Service Value Chain

Plan

- Understand current and future skills requirements, and staff turnover

Improve

- Continually adapt to meet evolving business needs

Engage

- Understand & forecast changing demand for services and how this will impact workforce

Design and transition

- Understand competences needed for Agile, DevOps, etc. define training plans

Obtain/build

- Training, mentoring, succession planning, recruiting or sourcing skills

Deliver and support

- Measure how knowledge, abilities and attitude impact practices

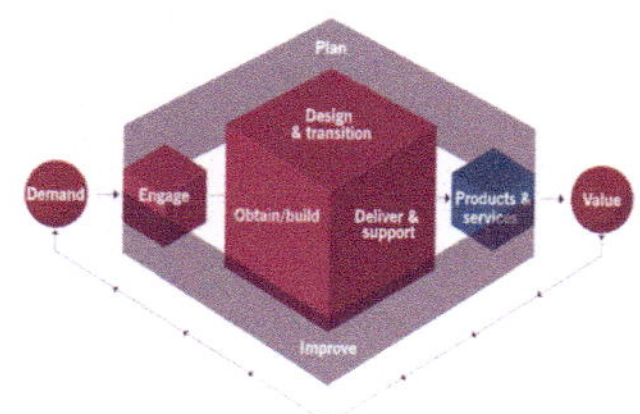

Role vs Job

Key message

- A role is a set of responsibilities, activities and authorizations granted to a person or team, in a specific context
- A job is a position within an organization that is assigned to a specific person
- A single person may, as part of their job, fulfil many different roles. A single role may be contributed to by several persons

Results based measuring & reporting

Why measure?
– Justify, Direct, Validate, Intervene...

Note: Differentiate Output & Outcome!

Examples of why we measure:

- To identify current and planned future state
- To measure achievement of improvements, changes, or plans
- To measure progress towards goals or objectives
- To support business decisions
- To drive behaviours
- To understand how well services are meeting customer needs/expectations
- To identify opportunities for improvement

Measuring behaviour or results

When should behaviours be measured?

- There is no clear relationship between behaviours and results
- If outcomes are too far in the future
- If results are not in control of those we are measuring

When should results be measured?

- There is a clear link from behaviours to results
- Easiest way to translate strategic objectives into measured actions
- Good when people have skills and abilities needed to complete their work and can correct their own behaviours
- Motivates people to improve, allows autonomy on how to deliver the results

Performance goals

How to set and measure individual performance goals:

- Establish separate goals in face-to-face meetings for each organization, function, team, and individual
- Ensure that the goals are measurable and documented and make it easy for all to track their progress
- Express the goals in specific terms
- Adapt the goals to the individual
- Adjust any goals that prove to be unrealistic

Measuring Team performance:

- Ensure that the team's goals are aligned with the organization's goals
- Begin with the team's customers and the work process that the team follows
- Measure both team and individual performance
- Qualitative and quantitative measures are needed
- Measures must continue to change and evolve to ensure that there are changes in behaviour to drive continual improvement

Summary

Organizational structures are central and clearly impacts an organizations ITSM capabilities

Service Management requires a broad set of business, communications and industry skills for a practitioner to be fully effective. Generic skills should be understood and defined beyond technical and process capabilities

Both cooperation and collaboration are effective forms of teamwork; it is important to understand and exploit their specifics. Building a strong team culture is important in order to have functional and effective team

IT Service Management requires good relationships and communications with all stakeholders, internal and external

Why do we need to prioritize?

It is important to remember that queues can occur wherever the demand for work exceeds the capacity to complete work within the expected time. In an ideal situation, an organization would have no variation in demand and would have the appropriate quality and quantity of resources needed to satisfy it. However, organizations often face varying demand for services alongside a fixed capacity. This imbalance between demand and capacity creates queues or backlogs in which work items need to be prioritized.

Prioritization is a method commonly associated with support and software development work; for example, prioritizing incident records or prioritizing a sprint backlog, but it is actually universal in scope

Definition: Prioritization is the method of managing work according to relative importance, by creating an ordered sequence of work to fulfil more important demands first

Managing work as tickets

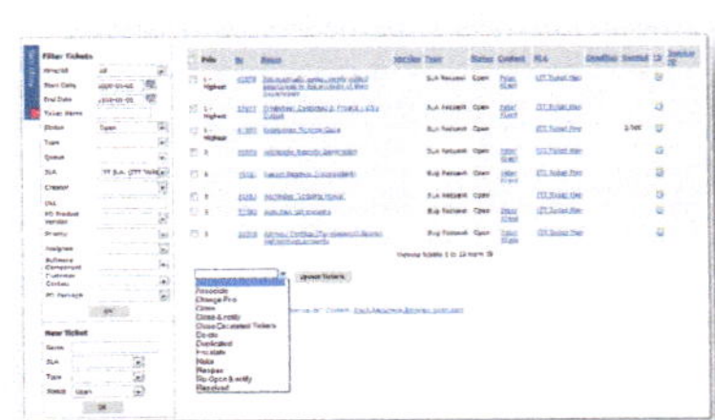

A ticket is a record of work. It is important to recognize that the ticket represents a discrete unit of work, and its current state within its expected lifespan

A busy service provider organisation performs many different tasks and activities simultaneously, so it is vital that they have the means to record and track their work. Without this, the risk of chaos, data and information loss, and a lack of accountability is high; all of these have significant consequences to the risks management and overall governance of the organization

Hence, tickets are important. They enable prioritization, communicate the current state of any given task to anyone who should know it, and enable high-value behaviours. Even after work has been completed, the archive of recorded tickets can continue to provide value as a source of data for management reporting and analytics

Tickets and design considerations

However, the user experience and behaviours that have built up around tickets can cause problems

Tickets do not cause queues to happen: ticket queues are not a fundamental property of a ticket. Rather, queues are the result of the way the organization allocates people and other resources to manage tickets

Hence, effective service design does not require tickets to be avoided or eliminated. They should simply not be the dominant influence on the user's experience. Design-thinking principles are crucial; they enable the service designer to focus on the stakeholders' specific challenges and identify user-focused solutions

Prioritization and demand management

The prioritization of work to create, deliver, and support services is necessary to co-create value while minimizing the costs and risks that arise from unfulfilled demand and idle capacity

At a high level, if demand is created when idle capacity exists, then there is no need for the prioritization of work; the work to fulfil that demand can start immediately

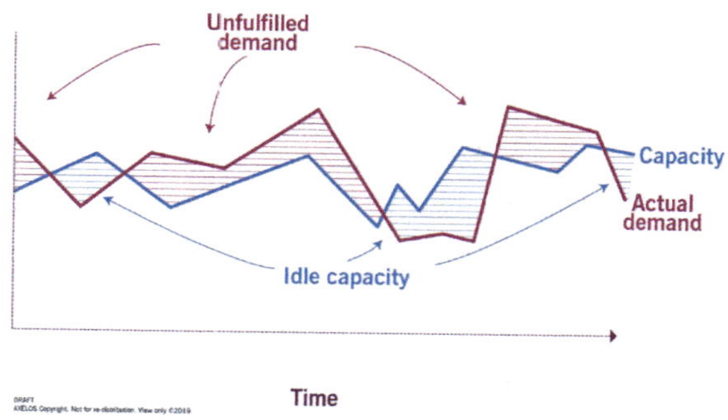

Figure 5.1 Demand variations and their effect on capacity

How to manage demand

However, when demand exceeds capacity, organizations often manage demand and minimize queues to avoid the need to prioritize work. Examples of this include:

- reducing variations in demand for value
- reducing variation in how much demand is taken into a value stream or step
- increasing how much demand can be satisfied within a given period of time
- reducing the cost of increasing or decreasing capacity
- using Shift-left techniques to deflect demand or prevent demand from being created

The above list highlights common methods of managing demand but, depending on context and complexity, other methods might be more appropriate

The Shift left approach

Shift left is an approach that moves work to the optimum place, closer to the source, or customer. This improves efficiency, speed and customer experience

Shift left has application in all areas of service management – from development, to operations and support

Success with shift left requires buy in across organisations and clarity on expected outcomes and benefits

How to prioritize work

The prioritization of work can occur at different levels of granularity, with different implications on the wider system, and with different levels of impact on user or customer experience. For example:

Prioritization conducted at a **value stream level** increases the need to manage user or customer expectations, keep them engaged, and provide regular status updates. This is because, from their perspective, value realization is being delayed

Prioritization conducted at a **value stream step**, action, or task level can result in the flow of work being constrained to the point that resources waiting to be used later in the value stream remain idle or a build-up of work (in a queue or a backlog) forms earlier in the value stream. This phenomenon is commonly referred to as 'creating a bottleneck' or 'creating a constraint'

There are many different techniques for prioritizing work and thereby minimizing queues and wait times

Prioritisation is context-specific

It is important to note that prioritization techniques are context-specific; a technique that works for one environment or type of work might not work for another

It is also common to find more than one prioritization technique working in tandem. For example:

- In the Cost of Delay method, prioritization considers the economic impact or penalty over time
- In the Weighted Shortest Job First (WSJF) method, prioritization considers the Cost of Delay and the duration of the work
- A common technique to prioritize incidents is to consider impact (an economic factor) and urgency (a time factor)

Value streams

Value streams are a powerful tool in the design, development and transition of new services

Key to understand value streams is to recognise that:

- Value streams are simple, but not necessarily simplistic
- There are many different value streams and various types of work do not simply follow the same route
- Value streams can either represent a design or ideal pattern of activity, or reflect the actual, observable patterns of activity
- The same resources can appear in different parts of the value stream

Value streams describe **how** value is created, from demand to value realisation, often through the delivery of services & products

ITIL service value streams

The ITIL service value chain is made up of six archetypal value chain activities

A sequence of service value chain activities is referred to as an ITIL service value stream, or more simply as a value stream

Note: A value chain can:

- Mention one, some, or all value chain activities, depending on context
- Repeat value chain activities, depending on the work in progress

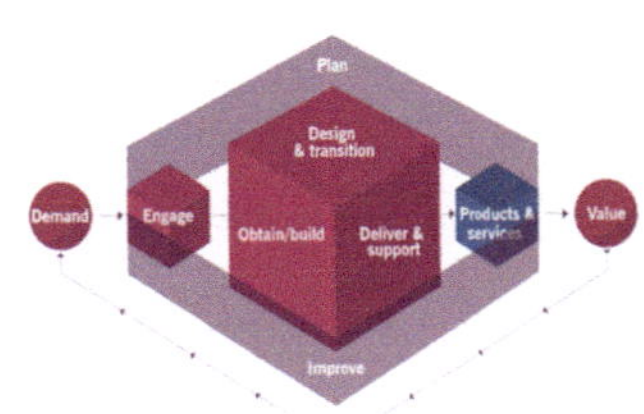

The ITIL service value chain is an operating model which covers the key activities required to effectively manage products and services

Value streams activity hierarchy

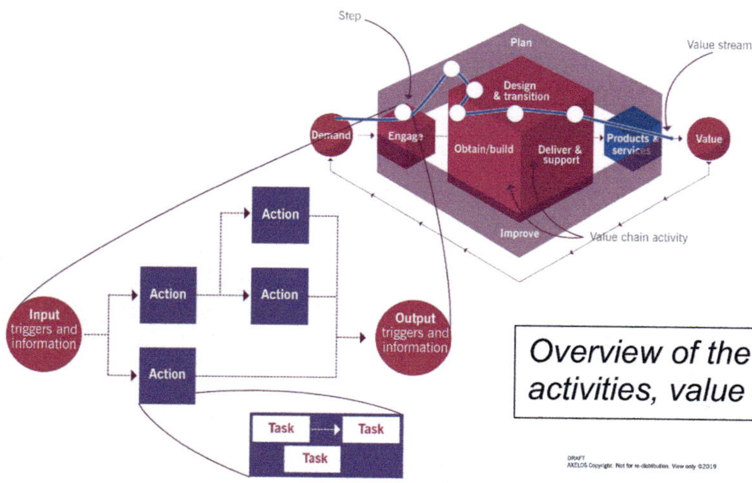

Overview of the relation between value chain activities, value streams, steps and tasks

Examples of value streams

A value stream will perform demand for value, for example:

- an incident generated by a monitoring tool
- a request submitted by a user
- end in the creation or restoration of value through functioning products or services
- use information provided either by:
  - one or more stakeholders (external to the organization executing the step or action) as inputs
  - another value stream
- use resources of the organization's IT and service management practices to generate required outputs
- use resources provided by consumers to generate required outputs
- often use some resources to modify other resources, in order to generate required outputs
- work within constraints and policies set either by management or governance system(s), or by external factors

Processes and value streams

Key differences between value streams and processes have to do with their focuses and how they are used. Value streams are focused around the flow of activity from demand or opportunity to customer value

So, process taxonomy, and management tools and techniques are applicable to value streams, however many processes are not value streams

Definition: A process is a set of interrelated or interacting activities that transform inputs into outputs. A process takes one or more defined inputs and turns them into defined outputs. Processes define the sequence of actions and their dependencies.

Definition: A value stream is a series of steps an organisation undertakes to create and deliver products and services to consumers

Value stream considerations

- Value streams can be documented from either of two perspectives maintained throughout the value stream:
  - aspirations
  - observed behaviour
- Value streams always starts with demand, and always ends with value being created or restored for one or more stakeholder. Thus, it is highly desirable to maintain an outside-in tone or language when documenting the value stream
- Value streams can loop around and repeat value chain activities, reflecting the context and the environment in which work is performed. However, a value stream can be as flexible as the organization needs it to be
- Value streams are a representation of work, as viewed at a certain level of granularity. Regardless of how the work is represented, it is critical that the level of granularity is uniform across the entire value stream
- The level of detail that would be represented in the value stream needs to be considered. The organization will need to decide whether the value stream will show details of all the actions, or provide an overview of the work
- Deciding what constitutes a separate step in a value stream, and what should be included within an existing step is key. Activities that are carried out by different teams are usually best shown in different steps. This can be very helpful for understanding where work is delayed by queues. The inclusion of multiple value chain activities also affect the value stream

Designing a service value stream

1) Define the use case, or scenario, for the value stream, by describing:
    a) The demand (preferably in non-technical terms)
    b) The trigger created by the demand
    c) The outcomes created by the value stream
    d) Value in the context of the value stream (value can be created or restored)
2) Document the steps required to traverse the service value chain from demand to value
3) Map the steps from no. 2 above to the service value chain
4) If necessary, decompose steps into actions and tasks
5) Identify practices and associated resources contributing to the successful completion of each step, action, or task including:
    a) Operational and management teams, or individuals
    b) Tools and technology
    c) Information and data (records, forms, or other artefacts)
    d) Any partners and suppliers that create outputs or outcomes using their own resources

The 5 steps above should be completed in a collaborative way (for example, a series of meetings or workshops). At the first stage of documentation, it is not necessary to get all the details correct, but to establish a broad and general understanding (or a "baseline") of the work needed to respond to demand and generate value

Describing a step

When describing a step in a value stream, the questions that need to be answered include:

- Name of the step
- The input trigger(s)
- Information needed
- Practice contributions
- Actions and tasks
- Constraints
- Outputs
- Estimated or Target Lead Time

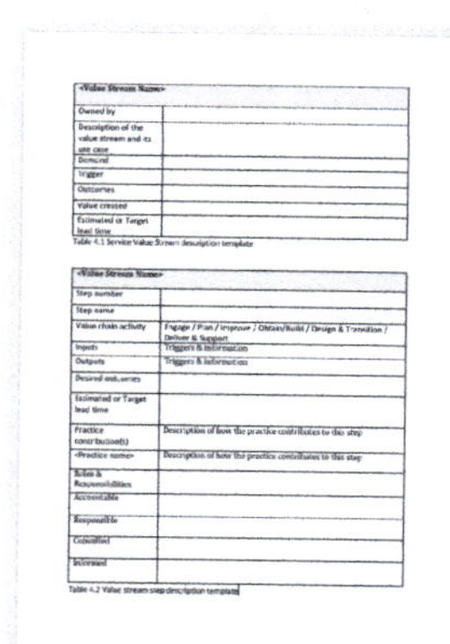

Value stream mapping

Value stream mapping helps organizations by:

- Visualizing more than just the single-process level, i.e. assembly, welding, etc., in production. The flow from opportunity to value is visible and clear
- Making the sources of waste in each value stream more visible
- Providing a common language for talking about value streams and processes
- Making decisions needed about the flow apparent, so they can be discussed. Otherwise, many decisions on the work floor just happen by default
- Tying together Lean concepts and techniques, which discourages using just one or two concepts or techniques in isolation
- Forming the basis of an implementation or improvement plan. By helping organizations design how the end-to-end flow should operate, value stream maps become a blueprint for implementation
- Demonstrating the linkage between the information flow and the material flow

Key metrics for value streams

There are several important metrics which can be defined at the value stream, step, action, and task level:

1. **Cycle Time**: This is the amount of time required to complete a discrete unit of work, converting input(s) into output(s)
2. **Wait Time**: This is the amount of time a discrete unit of work waits in a queue before it is worked on
3. **Queue**: The number of discrete units of work waiting to be operated on by the step, action, or task
4. **Lead Time**: This is the sum of Cycle Time(s) and Wait Time(s) from start to finish
5. **Work in Progress**: The number of discrete units of work currently being operated on, but which are not yet completed
6. **Throughput**: The rate at which work enters or exits the system

Different levels of representation

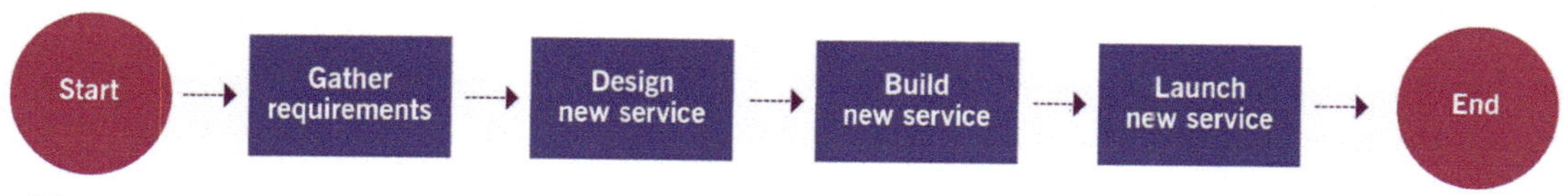

Figure 4.3 Simple representation of a value stream

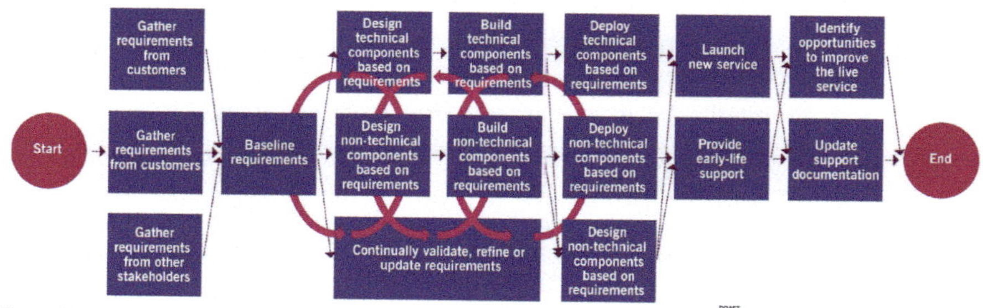

Figure 4.4 Complex representation of a value stream

Model value streams in ITIL 4

ITIL 4 includes two common value stream models that cover:

- **Development of a new service**
- **Restoration of a live service**

These can be seen as a good starting point, but should be adapted to the specific needs of the organisation. The templates previously introduced are used, but can be tailored by using additional information. An example is the "practice contribution" field that can include and consider any or all of the four dimensions of service management:

- **Organizations and people:** Skills, management authority, funding, staffing, etc.
- **Information and technology:** Tools, databases, data objects, information objects, etc.
- **Partners and suppliers:** Vendors that provide products and services to the organization, etc.
- **Value streams and processes:** Processes, procedures, templates, etc.

Development of a new service

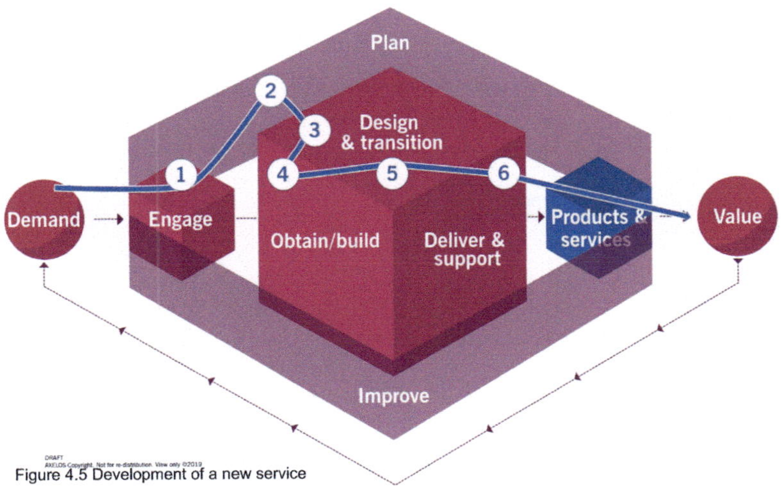

Figure 4.5 Development of a new service

Step 1 - Understand and document service requirements

Step 2 - Decide whether to invest in the new service

Step 3 - Design & architect the new service to meet customer requirements

Step 4 - Build, configure, or buy service components

Step 5 - Deploy service components in preparation for launch

Step 6 - Release new service to customers and users

Development of a new service

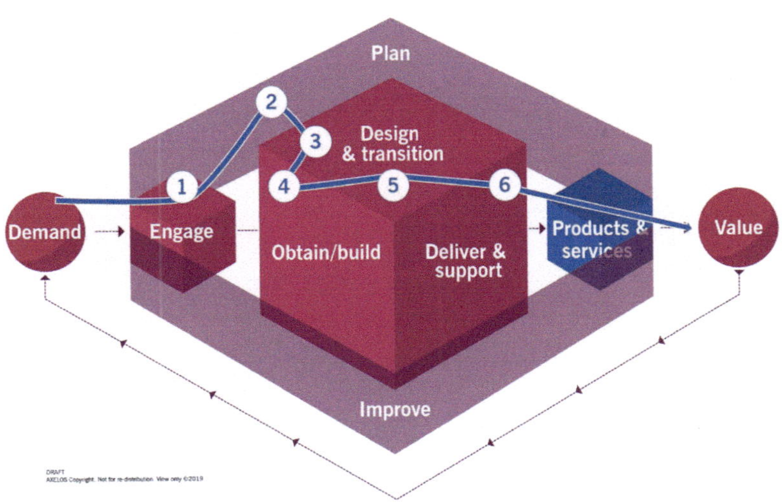

Demand and value
- Demand triggered..

Development of a new service

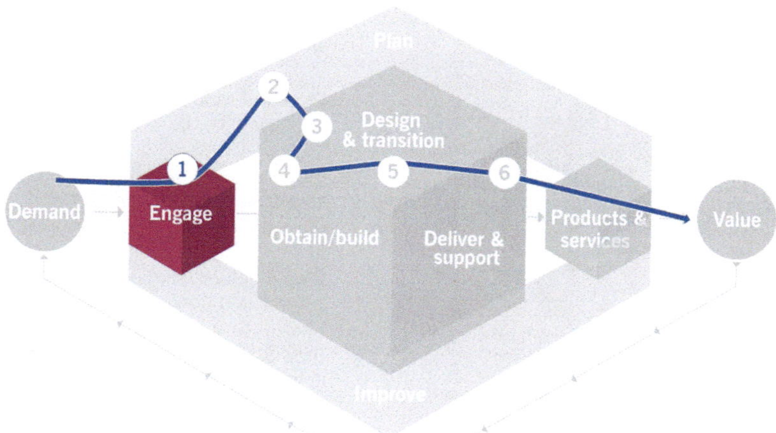

Acknowledging and documenting the request

Development of a new service

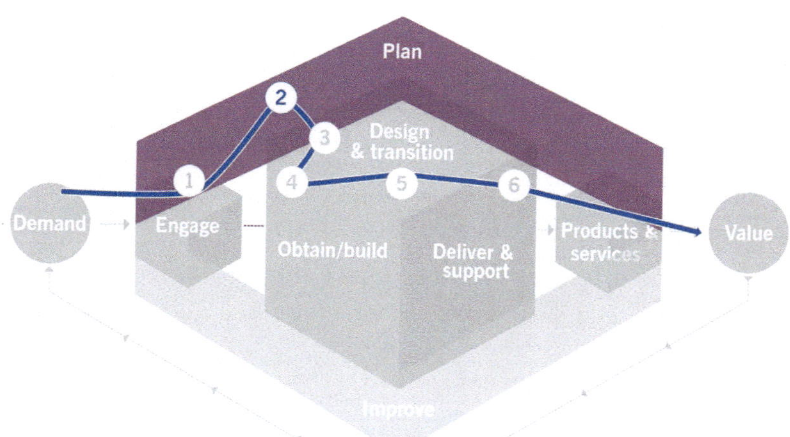

Assessing the requirements, and making a decision on investing in the new features

Development of a new service

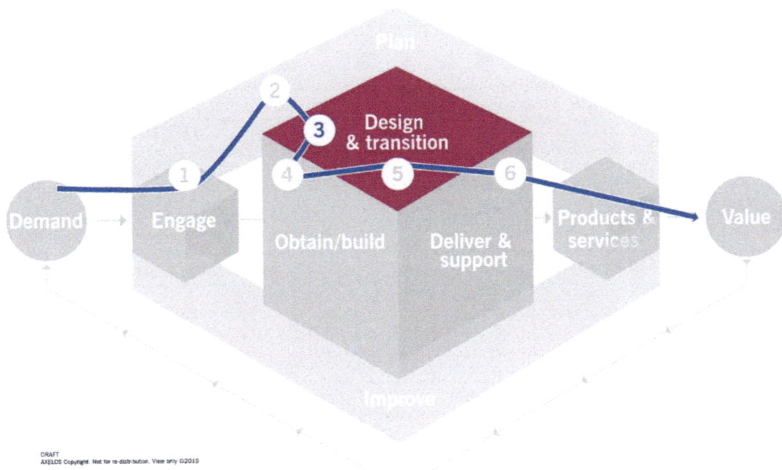

Designing the service to meet new requirements

Defining requirements for service

When defining requirements, an organization should reflect the needs of all relevant stakeholders

Requirements for service components should cover a broad range of topics and should not be limited to the functional needs that are articulated by users

A common approach to defining requirements is to focus on the technical (functional and non-functional) features of a product or the technical quality of a service. It is sometimes better to define requirements using outcomes instead

The MoSCoW technique relies on cooperation, and often negotiation, between all relevant stakeholders. As a result, defining and prioritizing requirements is often a complex and emotionally-charged exercise

Development of a new service

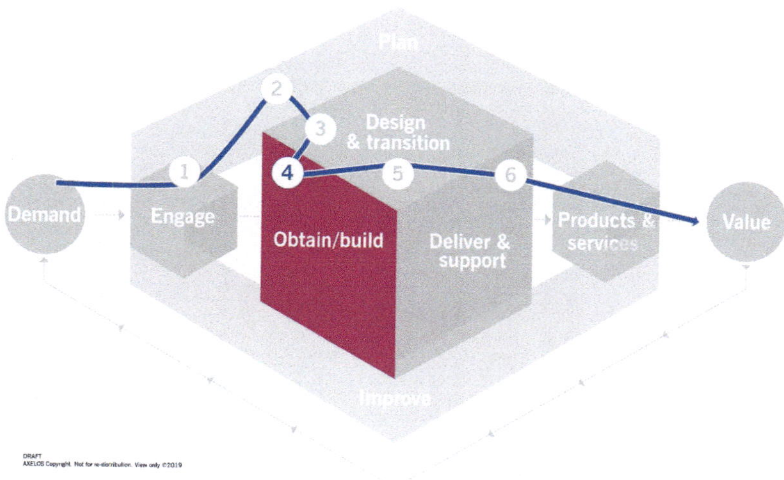

Creating service components and validating them against business needs

Defining requirements for service

When defining requirements, an organization should reflect the needs of all relevant stakeholders

Requirements for service components should cover a broad range of topics and should not be limited to the functional needs that are articulated by users

A common approach to defining requirements is to focus on the technical (functional and non-functional) features of a product or the technical quality of a service. It is sometimes better to define requirements using outcomes instead

The MoSCoW technique relies on cooperation, and often negotiation, between all relevant stakeholders. As a result, defining and prioritizing requirements is often a complex and emotionally-charged exercise

'Build vs buy' considerations

Building service components using existing resources works better in contexts where:

- the service component heavily relies on knowledge of the organization and its business
- customer demand for personalized products, services, or experience is high
- the ecosystem is volatile or subject to rapid change
- service components lack mass-market adoption
- compliance to standards and policies is a high priority
- the service provider is undergoing rapid growth or transformation, which can lead to inconsistent or changing requirements

Buying (or otherwise acquiring) service components from partners and suppliers works well when:

- in-house resources are scarce or highly utilized in other areas
- skills or competencies needed to create, operate, and maintain components are highly specialized and take time to build
- the processes to build products and services are immature and need to be developed and implemented
- components or services are highly commoditized
- the demand for service components is low or subject to significant fluctuation
- the service component is not core to the strategy, brand, or competitive differentiation of the service provider
- creating the service component is predictable and repetitive work
- the ecosystem is stable and generally not subject to volatility.

Commodification

When considering whether to build or buy a service component, the current level of 'commodification' and ongoing industry trends to commodify that component should be considered

For example:

- As the use of data centres became more prevalent and as computing power increased, virtualization tools emerged to manage virtual infrastructure
- As the cost of computing and storage fell and as virtualization tools matured, cloud computing models (infrastructure-as-a-service) emerged
- As cloud computing models matured, other cloud-enabled platforms emerged, such as platforms-as-a-service, software-as-a-service, and most recently, functions-as-a-service

This means that when considering whether to build or buy a service component, the current level of 'commodification' and ongoing industry trends should be considered

Sourcing models and options

A sourcing model is a component of an overall sourcing strategy. The selection of a particular sourcing model will reflect the organization's framework for managing, reporting, auditing, and ensuring compliance with the organizations vision, mission, ethics, and values across the organization's service supply chain

Common sourcing models include:

1. **Insourcing** - Where the organization's existing resources are leveraged to create, deliver, and support service components
2. **Outsourcing** - Where the organization transfers the responsibility for the delivery of specific outputs, outcomes, functions, or entire products or services to a vendor

Outsourcing models

Outsourcing models can be further subdivided based on where vendors or their resources are located. There are three categories of vendor location:

- **Onshoring** - Vendors are in the same country
- **Nearshoring** - Vendors are located a different country or continent, but there is a minimal difference in time zone
- **Offshoring** Vendors are located in a different country or continent, often several time zones away from the organization

When outsourcing work, organizational resources that remain after work has been shifted to the vendor is referred to as the 'retained organization'

Development of a new service

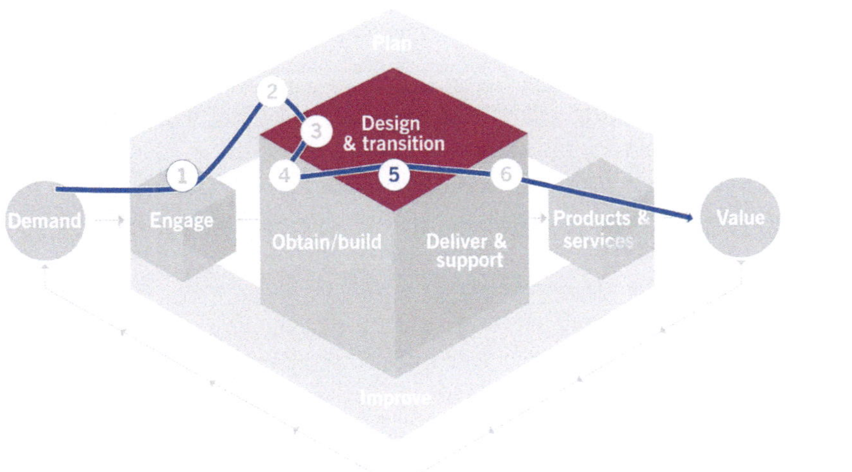

Modifying live products and services

Development of a new service

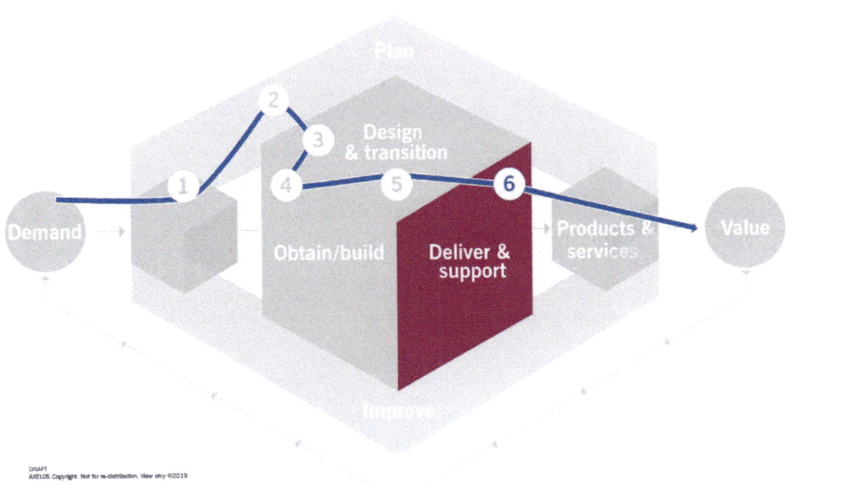

Making new features available to customers

Outsourcing success factors

Many organizations decide to outsource work to simply reduce short-term operational costs

A holistic approach to considering outsourcing will help to reduce the likelihood of these negative impacts. It is important to consider:

- whether it is important to retain knowledge and skills that could potentially be sent offshore
- what the impacts are to enterprise risk management when sending work offshore: what risks are mitigated, worsened, or created as a result?
- whether the industry or scope of work supports outsourcing
- what the cultural or language differences between the vendor and the organization are
- whether and how much management overhead will be added when outsourcing work

Restoration of a live service

Step 1 - Acknowledge and register the user query

Step 2 - Investigate the query, reclassify it as an incident, and attempt to fix it

Step 3 - Get a fix from the specialist team

Step 4 - Deploy the fix

Step 5 - Verify that the incident has been resolved

Step 6 - Request feedback from the user

Step 7 - Identify opportunities to improve the overall system (that is, service, value stream, or practices)

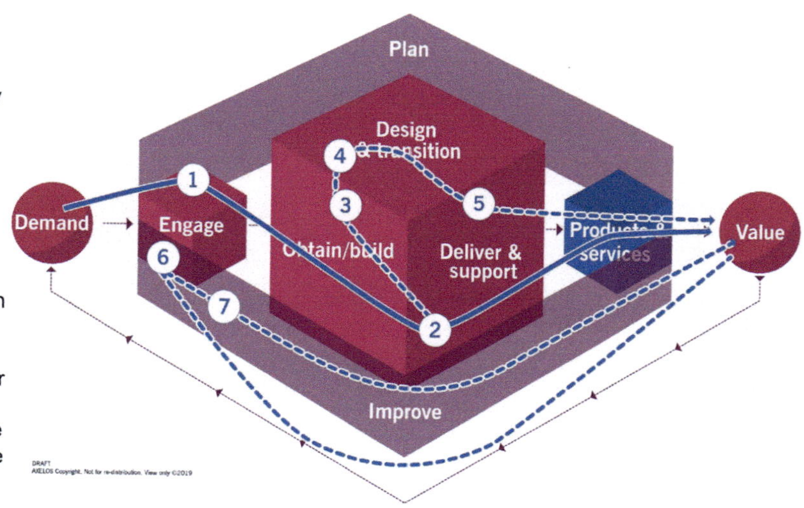

Restoration of a live service

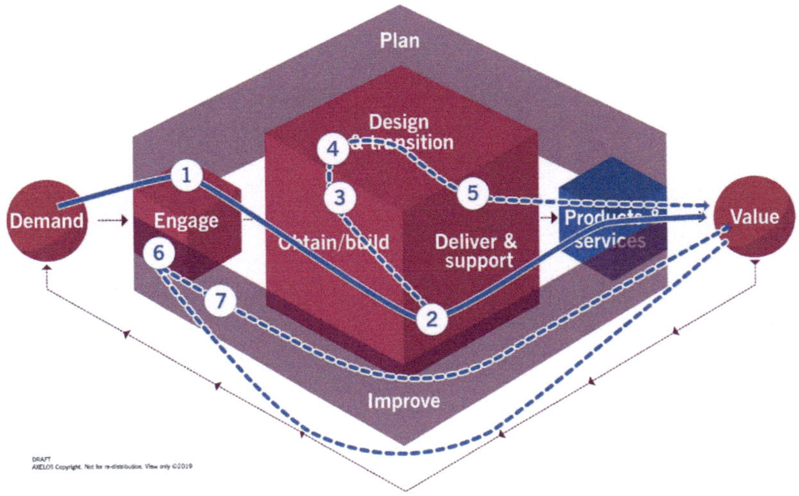

Restoration of a live service

Acknowledge and register the user query

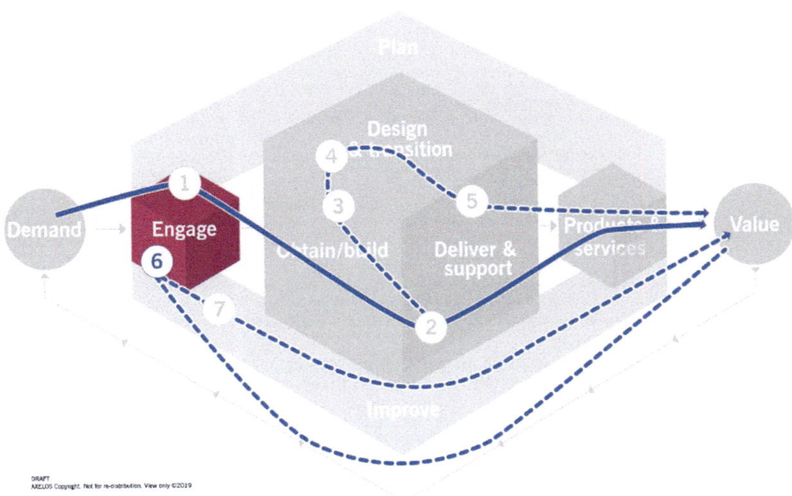

Restoration of a live service

Investigate the incident and attempt to fix it

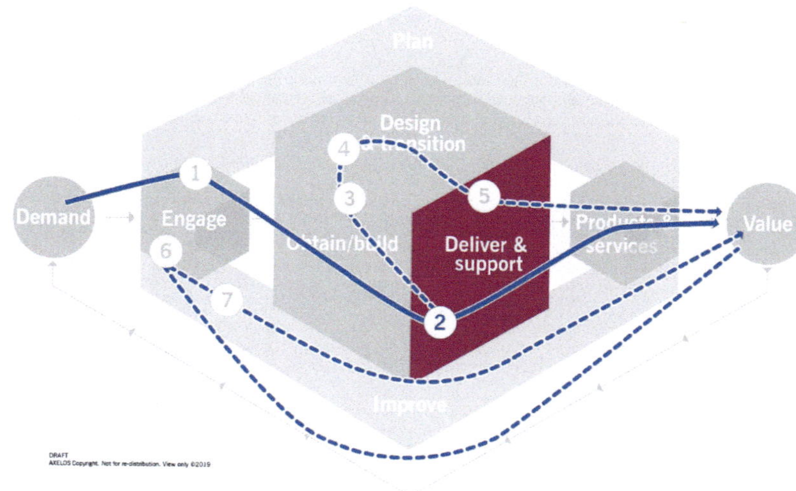

Restoration of a live service

Get a fix from the specialist team

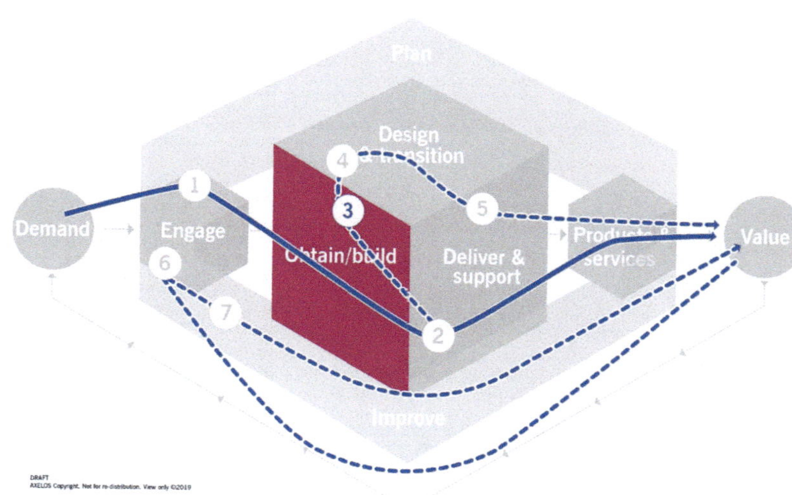

Restoration of a live service

Deploy the fix

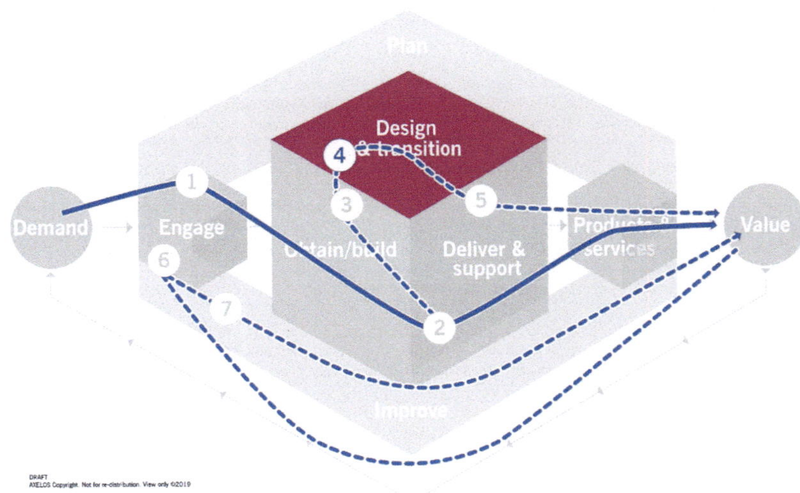

Restoration of a live service

Verify that the incident
has been resolved

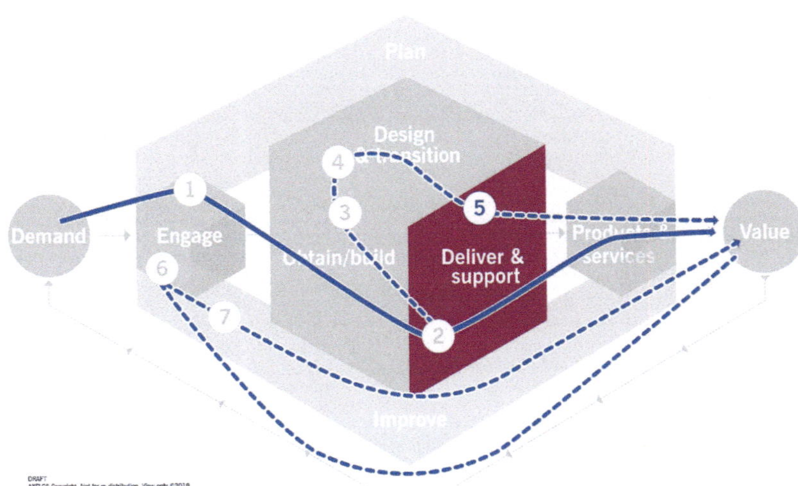

Restoration of a live service

Request feedback
from the customer

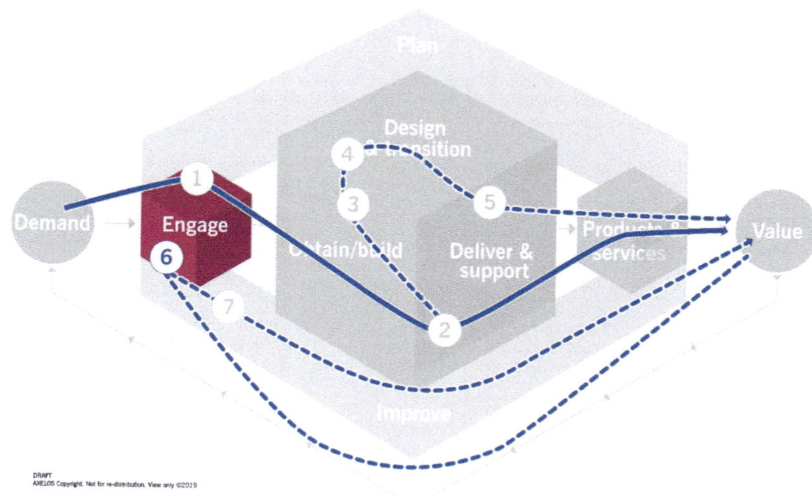

Restoration of a live service

Identify opportunities to
improve overall system

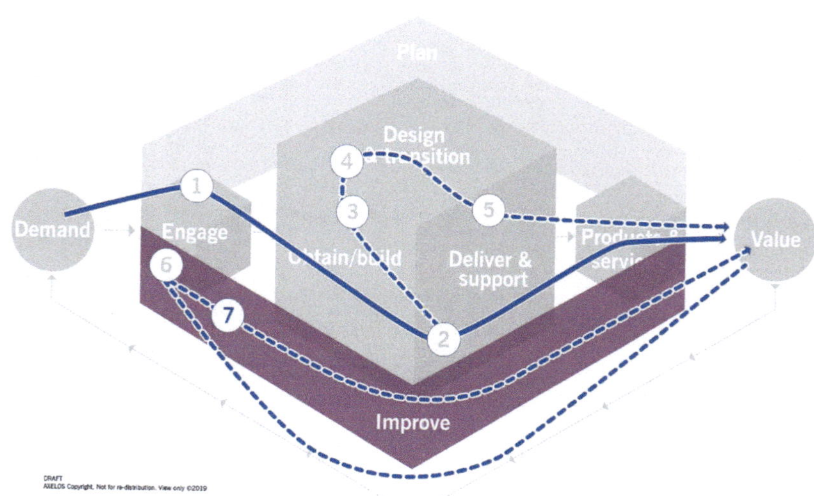

Summary

We have just talked about:

- ✓ This section has covered how to work with organisational structures and team culture to ensure organisational agility and resilience
- ✓ The section has also covered how to use Value streams with support of the ITIL practices, with special focus on the development of a new service
- ✓ Considerations around prioritization of work have been explored
- ✓ Key areas discussed:
  - *Organisation and organisational structures*
  - *Cooperation and collaboration*
  - *Roles, jobs and competencies*
  - *Service mindset and service empathy*
  - *Value streams & value stream mapping*
  - *The 6 key activities of the ITIL service value chain*
  - *Sourcing; outsourcing and insourcing*
- ✓ How the ITIL principles can be a good starting point

Q: Which is an example of team collaboration?

A. Working with others to achieve individual goals

B. Installing technology to facilitate communication among team members

C. Working together to achieve a shared organisational objective

D. Aligning individual and group goals and KPIs

Q: An organisation that has a hierarchical structure is reviewing its project management practice and is considering the various ways to introduce a new or c hanged service. Which is an advantage of an agile approach?

A. The ability to focus on increasing speed to market and making changes in line with business objectives

B. The option of using industry-standard project management models to deliver a phased, predictable project

C. The ability to focus on a clear go live date to which the various work streams can be aligned

D. The ability to maintain the organisation's current structure which would speed up decision-making

Q: Which concept is concerned with creating good working relationships with other people by considering their intellectual and emotional needs?

A. Employee satisfaction measurement

B. The value of positive communication

C. Organisational structure

D. Automated interactions

Q: An organisation intends to improve the quality of support of its services. It recognises that some ways of working are not focused on creating value. Which is an example of a working practice that the organisation should STOP?

A. Discussing requirements and expectations of service availability with customers to determine incident prioritisation categories

B. Conducting a survey to collect feedback from users on how they would like the interface of the self-service portal to look

C. Engaging with consumers to understand if the service support model will be fit for purpose following a planned organisational re-structure

A. Using the response and fix times from the organisation's hardware vendor to drive incident resolution times for customers

Course schedule

- ✓ Day 1: ITIL 4 Foundation
  - Key concepts, 4 dimensions of SM, ITIL Service Value System, the service value chain & 7 guiding principles
- ✓ Day 2: Create, Deliver and Support
  - How to use Value streams with support of ITIL practices
- Day 3: Drive Stakeholder Value
  - Customer journeys, relationships and service offerings
  - How to act together to ensure continual value co-creation
- Day 4: High Velocity IT
  - The Digital enterprise and how to use guiding principles
- Day 5: Deliver, Plan and Improve and Exam
  - Direction and Planning, GRC, Communication & OCM
  - Work with sample exam and examination

ITIL 4:
DRIVE STAKE-HOLDER VALUE

Key learning requirements

DRIVE STAKEHOLDER VALUE

- Understand how customer journeys are designed
- Know how to foster stakeholder relationship
- Know how to shape demand and define service offerings
- Know how to onboard and offboard customers and users
- Know how to act together to ensure continual value co-creation (service consumption / provisioning) Know how to realise and validate service value

Introduction

This publication aims to guide stakeholders, whether they are customers or service providers, through the principles and practices of co-creating value through services

This section of the course will focus on:

- The concept of customer journey
- The concepts mutual readiness and maturity
- How to foster stakeholder relationship
- How to onboard and offboard customers and users
- How to act together to ensure continual value co-creation
- How to realise and validate service value

The customer journey

Engagement is crucial to optimizing service value. This is because service value is always **co-created** by users, customers, sponsors, service providers, and any other parties involved in the service

To drive stakeholder value, all stakeholders must contribute to the co-creation of service value. This section will discuss the main steps of a **customer journey**, providing guidance on how to co-create the most valuable journey

The customer journey can be broken down into seven steps:

The customer journey – 7 steps

1. Explore Understand markets and stakeholders	The customer journey often begins before the service provider and service consumer have established a relationship. Both parties may explore their own needs and market opportunities for identifying partners that may contribute to the fulfillment of their respective needs. This exploration may include aspects such as operational context, strategic objectives, and organizational capabilities.
2. Engage Foster relationships	An important precondition for co-creating value through services is a functioning relationship between the service provider, service consumer, and other stakeholders. Good relations are a prerequisite for a cooperative relationship or partnership.
3. Offer Shape demand and service offerings	To determine whether the parties may benefit from a mutual service relationship, the service consumer and service provider should build a business case and articulate, shape, and match their demand and supply in the form of requirements and service offerings. Products and services can only be designed when the service consumer needs are well articulated and understood.
4. Agree Align expectations and agree service	It is crucial to align expectations, plan value co-creation and tracking, and agree service scope and quality before investing.
5. Onboard Get on board or leave the journey	Whether the parties came to an agreement or not, they must undergo a transition that involves the integration or separation of both parties' resources.
6. Co-create Provide and consume	The service consumer makes use of accessible service provider resources, consumes the goods provided, and acts together with the service provider to co-create value based on the agreed service offerings.
7. Realise Capture value and improve	Value must be tracked and driven based on the value co-creation planning and improvements must be applied to continually maintain and increase the service value.

Table 1.1 The steps of the customer journey

Understanding the customer journey

The customer journey is not only relevant for the service consumer. The service provider must also identify, understand, and master the customer journey

The purposes of identifying, understanding, and mastering the customer journey	For the service consumer	For the service provider
Facilitate outcome and experience	• To gain optimal service value and experience from the service relationship • To understand what the service consumer needs and desires, not just what the customer states	• To identify and support specific service consumer behaviours and outcomes • To optimize and improve products, services, and customer journeys for future value realization
Optimize risk and compliance	• To ensure key service consumer risks have been identified and addressed	• To focus on the customer satisfaction issues and key areas with the highest payouts related to costs
Optimize resources and minimize costs	• To work together with the service provider to commit and optimize the use of resources during the service lifecycle	• To work together with the service consumer to commit and optimize the use of resources during the service lifecycle • To be fair and transparent regarding costs

Table 2.1 The purposes of identifying, understanding, and mastering the customer journey

Stakeholder aspirations

It is important to understand stakeholder value, which requires exploring the functional, social, and emotional dimensions that explain why stakeholders make certain choices in order to understand why stakeholders need certain products and services

The *Disney Institute compass* includes four considerations:

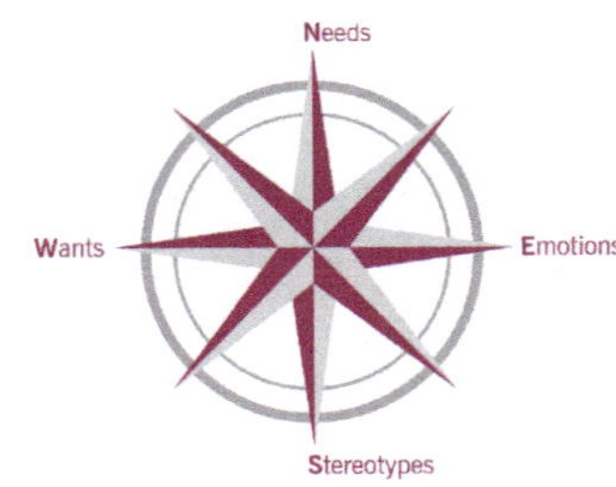

Needs - The basic reasons causing the stakeholder to begin the journey. Needs define the outcomes that are relevant to a stakeholder

Wants - Less definitive than needs, these are the things stakeholders would prefer to have or that would improve their experiences. Needs are requirements, but fulfilling wants is how service providers exceed expectations

Stereotypes - The preconceived notions, positive or negative, that stakeholders have about the experience. It is important to consider aspects such as history and culture when designing the journey

Emotions - Emotional intelligence and behavioural psychology are keys to understanding and mastering the emotional aspect of the journey because a stakeholder's emotional state will drastically affect their experience

Designing an end-to-end customer journey

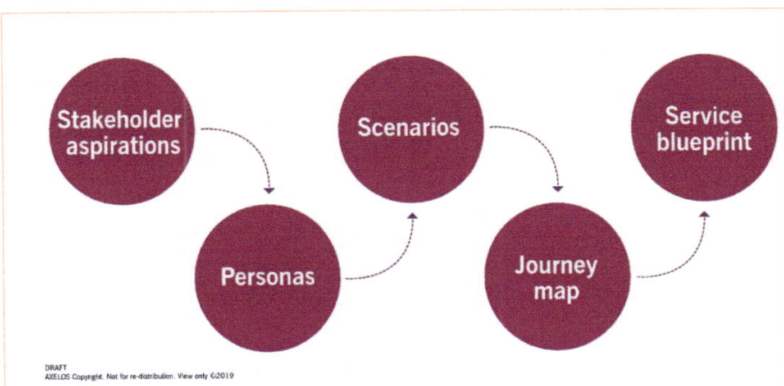

Figure 2.2 The stages involved in designing end-to-end customer journeys and experiences

Once the drivers are understood and a stakeholder-experience aspiration has been defined, the next stage is to identify, understand, and master the end-to-end experience of the customer journey by defining personas, scenarios, journey maps, and service blueprints

Touchpoints & service interactions

Service interactions include:

- Provision of access to resources
- Interaction with operant service provider resources
- Joint service actions

Definition: Service interaction
A value co-creating reciprocal action between a service provider and a service consumer

Definition: Touchpoint
Any event where a service consumer or potential service consumer has an encounter with service provider and/or its resources

Customer journey maps

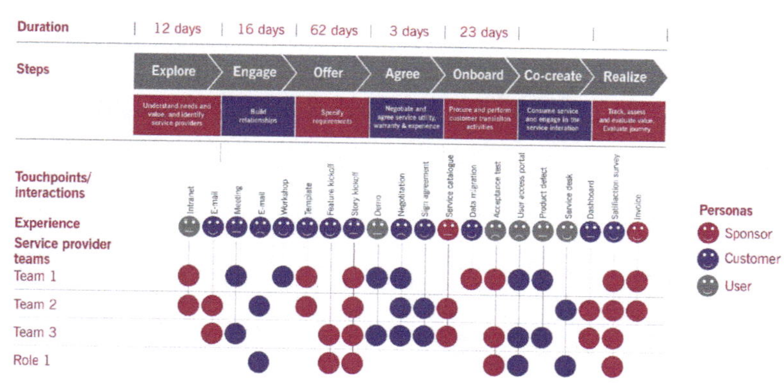

Figure 2.4 Example customer journey map

There are many ways to visualize the customer journey.

Simple maps typically include the steps of the journey, duration, touchpoints and service interactions, personas, service experience, and service provider teams and roles who interact with the service consumer

Step 2: Engage

Managing suppliers and partners
Building service relationship
Understanding service relationship types
Communicating and collaborating

The purpose of the engage step is to build transparency, continual engagement, and trust between the stakeholders in order to ensure a good mutual understanding of each stakeholder's preferences and experience

The service relationship model

In a service relationship, organizations will adopt the roles of service providers or service consumers. The two roles are not exclusive; organizations typically provide and consume a number of services at any given time

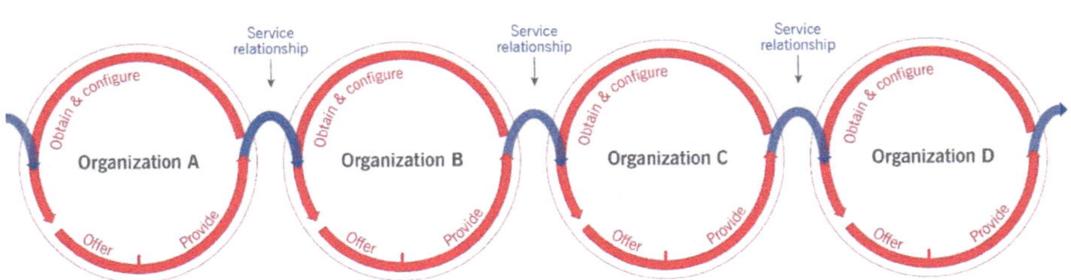

Figure 1.3 The service relationship model

Three types of relationship

Basic relationship

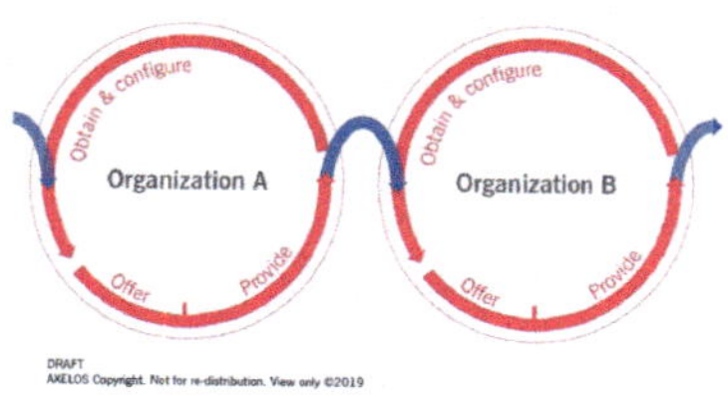

In a **basic relationship**, the customer may check the past performance of the service provider

Cooperative relationship

In a **cooperative relationship**, readiness to collaborate and communication mechanisms also become highly important when compared to basic relationships

Partnership

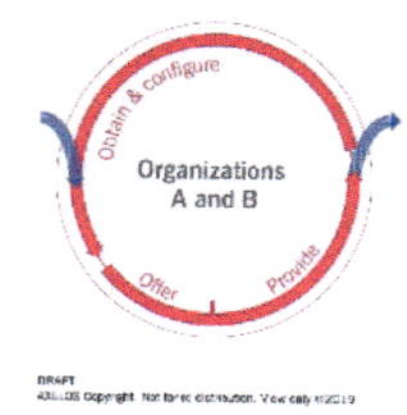

In a **partnership**, openness and trust are the key factors of mutual success. Readiness to collaborate becomes crucial

Engaging & fostering a relationship in different environments

	Basic relationship	Cooperative relationship	Partnership
Relationship maturity	• Ad hoc, order taker • Frequent misperceptions build distrust • The service provider is reactive and does not challenge customer requests • There is a lack of quality data to support cost or value analyses.	• Service provider, trusted advisor • There is a mutual understanding and appreciation of demand and supply • The service portfolio is appropriate to service consumer needs • The service provider engages early and often in the customer decision cycle, and there is a shared understanding of product and service value	• Strategic partner • The service provider and service consumer share common goals with a focus on value realization • There is clear accountability for achieving value from investments in products and services and quality data to support value analysis.
Approach for building relationship	• Loudest in, first out. • Frequent misperceptions may lead to distrust and reactive course changes, costs are usually transparent, but value may be hidden	• The routine is routine, but in innovation is a challenge • Collaboration is based on mutual respect and understanding • Portfolio is aligned to service consumer demand	• Shared goals for maximizing value, shared risks, and reward
Key attributes	• Little information sharing • Single channel of communication (dependent on single point of contact) • Driven by price • Easy to exit (usually a lot of alternative options)	• Service provider seeking opportunities to add value • Many points of contact • Forecasting, not joint planning • Limited information sharing • Expensive and resource intensive	• Both acknowledge the importance of each other • Streamlined process • Wide range of jointly executed activities • Free exchange of sensitive information • Difficult to exit

Table 4.3 Engaging and fostering relationship in different environments

Basic relationship – Advantages and disadvantages

Good for standard products and services, when the efficiency of operation is a cornerstone
A service provider in such relationship is interested in resilient and repetitive operation enabling the achievement of certain service levels with minimum effort and deviation

	Advantages	Disadvantages
Service consumer	• Easy to exit • Easy to control	• Emphasis is placed on efficiency and transactions • Hard to develop a deeper relationship • Hard to assess service value
Service provider	• Single channel of communication • Easy to measure and report • Building scale and operational • efficiencies into service management practices • Standard customer management approach	• Emphasis is placed on efficiency and transactions • Difficult to develop a trustworthy relationship • Little to no information sharing • Easy for the customer to exit • Customer is driven by price • Little opportunity to deliver differentiated customer experiences

Table 4.4 Pros and cons for a basic service relationship

Cooperative relationship - Advantages and disadvantages

The service provider usually tailors the products and services to the service consumer needs. Customers expects that providers will think about service outcome and experience, not only service levels. Service providers are expected to look for new opportunities to create additional value for the service consumer

	Advantages	Disadvantages
Service consumer	• The service provider tailors the service to specific service consumer needs • New opportunities to do more business with the service provider	• Mutual activities may feel uncoordinated • Expensive and resource intensive
Service provider	• Higher service consumer dependency on the service provider • More information is received from the customer, opportunity to be far more effective in helping to drive valuable solutions	• Mutual activities may feel uncoordinated • Expensive and resource intensive • Higher risk of mistakenly allocated resources • New operational complexities emerge • Customer brings internal problems to the service provider • Less efficiency and control over the customers

Table 4.5 Pros and cons for a cooperative relationship

Partnership – Advantages and disadvantages

In a partnership, the service provider and the service consumer may act as one organization coordinating activities across a great range of functions and processes. As the level of interdependency and integration grows, both parties may align on a strategic level by setting goals and priorities together

	Advantages	Disadvantages
Service consumer	• Transparency allows both sides to identify inefficiencies and work together to fix them, which leads to mutual cost reductions • Long-term planning opens new market opportunities	• Lock-in may prevent the customer from increasing requirements or exiting • Separation is painful and time consuming
Service provider	• Transparency allows both sides to identify inefficiencies and then work together to fix them, which leads to mutual cost reductions • Long-term planning opens new market opportunities • Chance to attract bigger and more strategic customers	• Separation is painful and time consuming

Table 4.6 Pros and cons of a partnership

The service relationship ladder

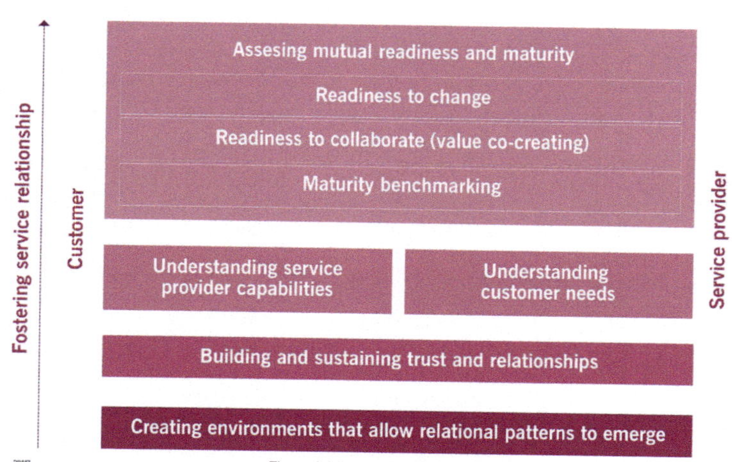

Figure 4.3 The service relationship ladder

The relationship management practice is applied to the customer journey through a service relationship ladder, which covers the initial phase of a relationship

The other steps of the customer journey play a role in fostering and shaping the service relationship

The steps of the service relationship ladder

Step	Description	ITIL management practices and tools
Creating environments that allow relational patterns to emerge	• Before any communication and collaboration starts, a service provider and a customer need to have an encounter • The service provider should establish or take advantage of existing meeting places to bring the customer close enough to engage with and make mutual endeavours possible	• Relationship management - Managing communication channels - Providing points of contact - Marketing activities • Service catalogue management - Service catalogue as an invitation to conversation for the customer
Understanding service provider capabilities (performed simultaneously with step 4)	• When the customer has chosen the service provider, then the customer wants to ensure that the service provider has a proper set of capabilities available, to fully leverage the provider capabilities	
Understanding customer needs (performed simultaneously with step 3)	• Service relationship is mostly focused on satisfying service consumer needs • The service provider should leverage the relationship to uncover and understand the value that the customer wants to create	• Relationship management -Understanding customer needs activity -Business analysis
Assessing mutual readiness and maturity	• Service provider: Are we capable of value co-creation with the customer? Do our resources and practices match customer needs?	• BRMP (Business-Provider Maturity Model) • Maturity assessment and audits

Table 4.7 The steps of the service relationship ladder

Relationship management metaphors

The service relationship ladder can apply to different service relationships, but it must be adapted to specific systems

The business relationship management institute (BRM institute) has defined three different relationship management metaphors that are aligned with this approach:

- **Connector** - that facilitates productive connections, shapes demand/supply, and influences the stakeholders
- **Orchestrator** - that orchestrates roles, resources and capabilities, and coordinates and aggregates demand and supply
- **Navigator** - that facilitates convergence between the stakeholders, planning, and guides the involved roles

Initial engagement activities

In order to successfully engage with a service consumer, a service provider may guide a customer through the following stages

Step	Customer status	Customer activities	Service provider activities
Awareness	• Customer should be aware of the service/service provider	• Market research	• Marketing activities • Facilitating effective connections
Motivation	• Customer should be motivated to start a service relationship with service provider	• None	• Marketing activities • Understanding alternatives • Stimulate demand
Contacting	• It should be easy for the customer to start service relationship -Where to go, single point of contact -What is offered	• Contacting service provider • Browsing service catalogue	• Providing single point of contact • Managing service catalogue
Shaping expectation	• Service provider should make the customer expect good experience	• Check past performance and/or public ratings (if relevant) of service provider • Due diligence	• Aggregate and shape demand • Ensure appropriate capacity and capabilities for service provision

Table 4.8 Initial engagement activities

Initial engagement tools

The service catalogue supports service provider activities in the initial engagement with the service consumer and is a powerful tool for ensuring:

- the customers' awareness
- an initial invitation to conversation
- supporting discussions related to standard and non-standard service offerings

The service provider should work with the customer to provide clear information about choices and to outline which risks the customer is willing to accept

Another powerful tool used by service providers is customer relationship management (CRM) systems

Develop relational patterns

The service provider should create a comfortable environment that allows relational patterns to develop. For this, the following points should be considered:

- Interactions should be scheduled and plans adhered to even in situations where communication is unnecessary
- If the service provider provides points of contact but does not respond promptly, it can have a negative impact on the organisation
- Service incidents can lead to situations where the customer blames the service provider. In these cases, conflicts should be managed
- There is always some disturbance in the service provider organisation, customer organisation, or the market. The risk management practice can indicate which risks that threaten the cooperative situation are the most significant

Building & sustaining trust and relationships

Because the service provider and the customer must work together to co-create value, they need to build and manage a transparent relationship that supports mutual trust and focus on achieving the service outcomes while optimizing costs and risks

To do this effectively, the actions are usually referred to as disciplines, business relationship management and customer relationship management are used. In ITIL, both disciplines are covered in the relationship management practice guide

Trust may emerge in various ways, including:

- **Knowledge-based** - As time progresses, knowledge of the other group can increase the level of trust
- **Calculus-based** - In a service relationship, trust can build rapidly. In these situations, both groups can weigh potential opportunities against risks

Trust and relationship factors

- **The three C's trustworthiness model**

Hacker and Couturier have categorized trustworthy characteristics into three dimensions:

- **Capability** The ability to produce results
- **Commitment** The concern for common goals and other peoples success and welfare
- **Consistency** The ability to perform as expected in the same way

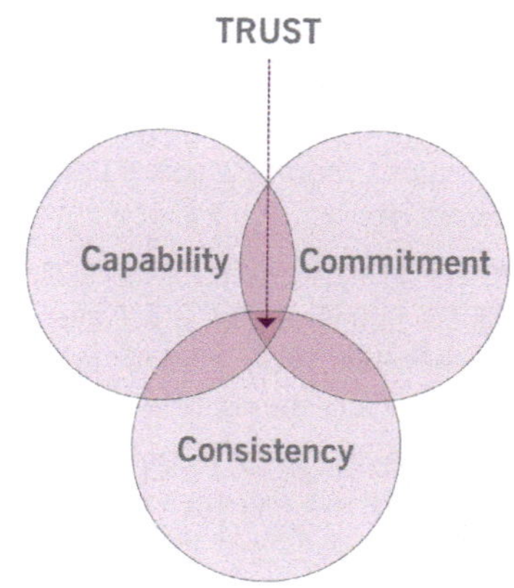

Figure 4.4 Three Cs trustworthiness model

The three C's model applied to a service relationship

Trust factors	Service provider	Customer
Capability	• Adequate knowledge and skills • Sufficient capacity for demand • Demonstrate agility/adaptability	• Demonstrate agility/adaptability
Commitment	• Show concern for the customer's success or share common/mutual goals • Be honest, respectful, and cooperative • Explain actions that impact the customer • Be familiar with the service consumer and its needs • Encourage/promote open, two-way communication	• Show concern for the service provider's success or share common/mutual goals • Be honest, respectful, and cooperative • Explain changes that impact the service provider • Encourage/promote open, two- way communication
Consistency	• Seek first to understand, then to be understood • deliver expected performance over time • Respond in a timely manner	• Seek first to understand, then to be understood • Disclose adequate amount of information

Table 4.9 The three Cs model applied to a service relationship

Building trust and relationship

Relationship type	Building trust and relationship	Customer	Service provider
Basic relationship	• Trust and relationships are built primarily through following • formal procedures and controls	• Share requirements with the service provider • Check past performance of service provider • Check public ratings and feedback from past and current customers • Due diligence/check • Evidence of compåliance to industry standards and/or certifications • Check SLA	• Manage service catalogue and service requests catalogue • Do not challenge requests from the customer • Satisfy demand/limit collaboration to the scope of the service catalogue • Provide reports Focus on outputs
Cooperative relationship	• Trust and relationships are built primarily through extensive communication and efforts of each party, made in order to achieve outcomes • Aligned outcomes • Feedback, semiformal controls	• Share needs • Be open to service providers proposals • Perform audit or maturity • Assessment cross references	• Manage and develop service portfolio according to changing customer needs • Advise on possible service solutions • Be driven by outcomes, not outputs
Partnership	• Trust and relationships are built primarily through sharing risks and rewards and focus on shared goals and value co- creation	• Establish open platforms for collaboration with the service provider • Share goals, risks, and rewards • Joint strategic/program/ projects planning • Demonstrate agility and adaptability to the • Changing environment	• Establish open platforms for collaboration with the customer • Demonstrate capability to enable consumer development and innovation • Focus on value realization in changing environment, not on fixed outcomes

Table 4.10 Relationship management activities

Continual building of trust and relationship

Numerous **factors threaten trust and relationships**, including:

- the natural tendency to separate both parties; for example, service providers from service consumers
- service provider's resources and practices constantly changing
- service consumer's risk profile and tolerance changing over time
- new employees in either party changing the nature of the service relationship (it is important to ensure timely engagement with newcomers)
- unavoidable customer complaints (formal customer complaints and escalation processes can mitigate this factor)

Understanding service provider capabilities checklist

The most popular ways to understand and evaluate service provider capabilities are through audits and maturity assessments

Organisation and people
• Understand organizational objectives and check whether the organization is structured to meet these objectives optimally
• Items/attributes to be assessed:
  • duplication (of activities, organizational functions, and so on.)
  • communication effectiveness (channels, frequency, feedback, and so on)
  • mode-of-working and decision-making structures (centralized versus decentralized, process-based versus hierarchical)
  • trust and transparency, level of control
  • knowledge and learning culture
  • Skills and competencies
	Information and technology
	• Assess quality of information used for decision-making and whether technologies support objectives and decision-making processes optimally
	• Items/attributes to be assessed:
	  • current technology projects
	  • integration
	  • technological debt
	  • technology trends (how technology
	  • developments may change the ways of
	  • working)
	  • reliability of data and information
	  • technical performance

Partners and suppliers • How effectively and efficiently does the service provider manage its partners and suppliers? • Does the service provider rely on strategic suppliers? • What is the possible impact to the service operation of key supplier shutdown? • Are they integrated in value chain seamlessly?	**Value streams and processes** • What processes are in place and how they function? • What is process performance? • How efficiently are value streams managed?

Table 4.11 Understanding service provider capabilities checklist

Understanding customer needs

Keep in mind, customers do not buy services; they buy the fulfilment of particular needs. They have jobs to be done

The service provider must understand these jobs in order to identify the service consumer's needs, preferences, and desired outcomes and experience

"People don't want quarter-inch drills. They want quarter-inch holes"
Theodore Levitt (1925 – 2006)

Value drivers

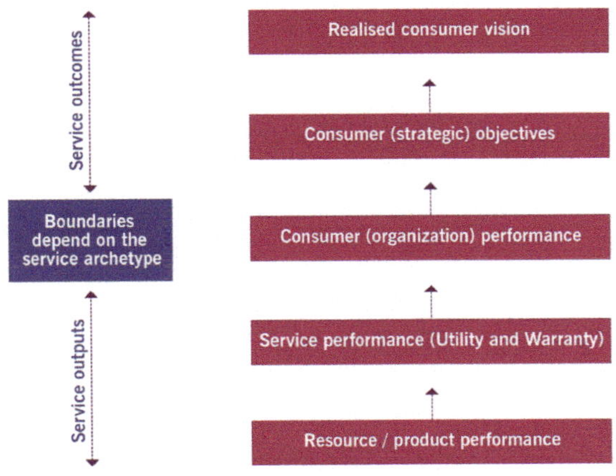

Figure 4.5 Service value driver framework

BRM institute guide defines two approaches on how outcomes may be linked to products and services, including:

- **Value-based approach (top-down)** The service provider discusses the customer's most pressing problems or objectives, then analyses initiatives, enablers, and technology
- **Solution-based approach (bottom-up)** The service provider discusses its products and services and seeks to connect them with a pressing consumer problem or objective, answering the same questions as in a value-based approach in reverse order

Risks and costs

Understanding the affected outcomes of a product and service, as well as costs and risks, should be part of discovering customer needs and building a service relationship

Customers perspective	Positive	Negative
Outcomes	**Supported outcomes**, including enhancing experience and meeting preferences • faster communication • single point of contact for all user requests • ticket tracking • user awareness of IT maintenance schedule	**Affected outcomes** • less direct communication between users and IT
Risks	**Risk removed** • lower response time • higher accuracy of service • request categorization	**Risk introduced** • single point of failure • higher dependency on users' • capabilities • sensitive incident data can be • reached by outsider
Costs	**Costs removed** • less service desk personnel	**Costs introduced** • Self-service administration and maintenance (in case of involving users into it)

Table 4.12 Positive and negative effects of a self-service portal

Key factors for service customer experience

Preferences influence the consumer's service experience, especially for mass-market services

Factor	Descriptions
Service quality aspects	• functionality, data and information, availability, performance, capacity, information security, continuity, accessibility, usability
Risks and compliance	• service complies with relevant regulations • service is aligned with consumer risk tolerance level
Price and costs	• perceived cost benefit ratio by the customer • service price compared to alternatives for example competitor's services
Design and convenience	• user experience • service streamlines an otherwise time-consuming process
Compatibility and interface	• architecture • compatibility with other services the customer is already using • availability across various channels or devices
Information, transparency and fairness	• all data about service pricing, actual usage, and so on, is easily available to the customer • customer can easily check data accuracy
Ability to control	• customer remains in control over service consumption and has tool to manage service options, pricing and so on
Social responsibility	• customer is assured that service provider conforms to applicable social responsibility norms, such as pollution, slavery, and fair trade

Table 4.13 Key product and service factors for service customer experience

Assessing mutual readiness and maturity

Mutual readiness is when both parties have completed appropriate checks and due diligence (i.e. audits) and built initial trust and are ready to form a working relationship in order to co-create value. Mutual readiness and maturity assessments may differ according to **relationship type**

- In a **basic relationship**, the customer may check the past performance of the service provider and refrain from investing in additional assurance measures such as audits
- In a **cooperative relationship**, the customer may assess the service provider maturity using audit and maturity assessment tools. Readiness to collaborate and communication mechanisms also become highly important when compared to basic relationships. It is crucial to align outcomes clearly across stakeholders and agree on feedback procedures
- In a **partnership**, openness and trust are the key factors of mutual success. Therefore, although formal capability maturity and past performance checks may occur, readiness to collaborate becomes crucial

Types of assessment (in the engage step)

The figure shows how relevant different types of assessment are to the different types of relationship

Attributes to assess	Basic relationship	Cooperative relationship	Partnership
Capabilities maturity and past performance (service provider)	Crucial	Moderate	Minor
Readiness to cooperate (both)	n/a	Moderate	Crucial
Readiness to change (customer)	n/a	Moderate	Crucial

Table 4.14 Types of assessment in the engage step

Relationship maturity

According to the business-provider maturity model, the maturity level, for example, basic, cooperative, and partnership, may be determined by a set of demand and supply characteristics

The business-provider maturity model	Demand	Supply
Basic relationship	• Transaction automation • Requests from business silos • Focus on functional performance	• Basic services • Custom/modified products for business silos • Focus on stability
Cooperative relationship	• End-to-end processes automation/business integration • Transactional data -> management information (information for decision making)	• Enterprise systems • Business processes improvement/reengineering • Open networks and gateways • Optimization/common infrastructure for a majority of IT services • Outsourcing commodity services
Partnership	• Innovation and growth • Business and market intelligence • Reconfigurableportfoliop of capabilities	• Technology-driven business capabilities • IT catalyst for market place differentiation • Standard interfaces

Table 4.15 The business-provider maturity model

Maturity assessment and benchmarking (checklist)

	Organisation and people	Value streams and processes	Information and technology	Partners and suppliers
Organisation and people	Are users ready to communicate with the service provider's agents? Can the service provider's agents understand users?	Are the service providers procedures modified to involve users? Can users follow procedures? Can we ensure that users follow procedures?	Are the service provider's products and services configured to provide the right level of access to the users? Are users trained to use technology	Are the service provider's supplier agents ready to communicate with users?
Value streams and processes	Can the service provider's agents follow the consumer's procedures? Can we ensure that agents follow procedures?	Which service consumer procedures provide input for the service provider's and vice versa? What are the required procedure for the service consumer and the service provider?	Do service provider products automate service consumer procedures correctly? Is there a need for change? Does the service provider know the service consumer's functional requirements?	Are the service provider's suppliers ready to follow service consumer procedures?
Information and technology	Do the service provider's agents have the skills and competencies needed to support consumer technology?	Are the service providers procedures covering technologies and information systems that the service provider needs to manage? (for example, change enablement, service configuration, management audit)	What information systems need to be integrated? What is the most effective and efficient way of integrating information systems? What changes in information systems are needed on both sides?	Do the service provider's suppliers have the skills and competencies needed to support the service consumer's technology?
Partners and suppliers	Has the communication channel between the service provider and the service consumer's suppliers been established? Are the service provider's agents ready to communicate with the service consumer's suppliers?	Are the service provider's procedures applied to the service consumer's suppliers? (for example, SIAM, change management, audit) Are the service consumer's suppliers ready to follow the service provider's procedures (if relevant)? How can we ensure it?	Are the service provider's products configured to provide the right level of access to the service consumer's suppliers?	Is the communication channel between the service provider's suppliers and the service consumer's suppliers established? Do the service providers ensure that all parties comply with all agreements and legal and regulatory requirements?

Table 4.16 The service provider and service consumer maturity assessment based on the four dimensions of service management

Assessing readiness for collaboration

Stakeholders are interested in collaboration because it is easier to achieve a goal with constant cooperation with other stakeholders. To assess readiness for collaboration, it is important to consider the following information (checklist):

Cooperative readiness factor	Checklist and references
• Has initial trust been built?	• The three Cs model
• Do we engage with relevant stakeholders? • Does management support cooperative activities?	• Stakeholder analysis • Stakeholders map (influence / interest grid)
• Do we create the organizational basis for close cooperation?	• Ensure alignment on principles and operating for activities • Check integrations and interfaces between information systems and other resources (four-by-four-dimensions matrix) • Communicate risk tolerance and constraints (risk register) Establish communication channels, especially for complaints (stakeholder communication plan) • Allocate resources (primarily, people, technology

Table 4.17 Readiness assessment checklist

Assessing organizational readiness for change

In some cases, successful service delivery may require the transformation of organizational practices and routines in order to obtain value from a service. This requires organizational change management. ITIL foundation defines the following attributes of an effective organizational change management practice:

Organization change readiness factor	Checklist
• ***Clear and relevant objectives***	• Is there a clear vision for the service relationship? • Is the reason for change effectively communicated?
• ***Strong and committed leadership***	• Are sponsors highly respected with successful track records? • Do the stakeholders trust management to consider their interests? • Are there any lateral leaders?
• ***Willing and prepared participants***	• Is there a plan for rewarding contributors? • Is there an adequate level of systemic knowledge within the organization? Do the stakeholders understand the full picture of the change and not only their own part?
• ***Sustained improvement***	• How much change resistant is within the organization? • Were past improvements successful? • How much change is currently taking place? • Is there a good balance between business as usual and innovation? If the organization overemphasizes responsibility for ongoing operations, innovation suffers. And if innovation is a sole focus, responsibility for current operational activities is ignored. • Therefore successful organizations should find the right balance.

Table 4.18 Organizational change readiness assessment checklist

Managing suppliers and partners

To some extent, every organization depends on services provided by other organizations. Therefore, relationships with suppliers and partners are equally important for the service provider and service consumer. This includes creating closer, more collaborative relationships with key suppliers to uncover and realize new value and reduce the risk of failure. The **service provider journey** includes the same steps as the customer journey:

- **Explore** understanding needs and value of identifying potential suppliers
- **Engage** building relationships with suppliers
- **Offer** shaping demand and specification of requirements
- **Agree** negotiating and agreeing terms and conditions for products and services provided by the supplier
- **Onboard** deciding whether products and services should be acquired from the supplier and performing transition activities
- **Co-create** the consumption of services and service interaction with the supplier
- **Realize** continually tracking, assessing, and evaluating value realization, as well as improving the whole journey

Service integration models

There are **four main models for service integration**, and each organization should consider what is the best model for them in order to transition to a more coordinated service-supplier landscape. The models are:

- **Retained organization as service integration and management** Where the retained organization manages all vendors and coordinates the service integration and management function itself
- **Single supplier** Where the vendor provides all services as well as the service integration and management function
- **Service guardian** Where a vendor provides the service integration and management function and one or more delivery functions, in addition to managing other vendors
- **Separate service integrator** Where a vendor provides the service integration and management function and manages all the other suppliers, even though the vendor does not deliver any services to the organization

Service provider as a service integrator

If a service provider acts a service integrator, it is usually responsible for building relationships and collaborating on behalf of the customer. To some extent, every service provider today is a service integrator

Step	Service integrator activities
• Creating environments that allow relational patterns to emerge	• Scan the consumer landscape in search for new suppliers and allow relational patterns to partners that may enable realization of the consumer strategy
• Building and sustaining trustand relationship	• The same as for a normal service relationship
• Understanding service providers capabilities	• Define criteria for managing suppliers based on relationship provider capabilities type, level of dependency, and risk. • Identify and categorize existing suppliers according to criteria and focus on the most important suppliers. • Assess supplier's capabilities according to the needs of the customer
• Understanding service customer needs	• The same as for a normal service relationship
• Assessing mutual readiness and maturity	• Track suppliers performance and compliance • Assess suppliers maturity • Assess the larger supply chain and manage risk related to the suppliers and their subcontractors influencing supplier's ability to deliver services

Table 4.19 Relationship management service integrator activities

The supplier management practice

The supplier management practice may be used to create a single point of visibility for all suppliers involved in service delivery to ensure consistency and enable value realization.

This helps to answer the following questions:

- Is the service provider effective and efficient when managing its partners and suppliers?
- Does the service provider rely on strategic suppliers?
- What is the possible impact of key supplier shutdown to the service operation?
- Are suppliers integrated in the service provider's value chain seamlessly?

The practice also ensures that agreements with suppliers are aligned with the expected service outcomes and customer needs and monitors their performance in order to ensure the terms, conditions, and targets have been met

Communicating and collaborating

Communication is fundamental for building relationships and trust. The effectiveness of communication depends on the type of relationship between those communicating

- **Cooperation** is working with others to achieve your own goals. These goals may form part of a common goal. However, there is a risk that the individuals/teams involved in the cooperation will work in silos. As a result, individual/team goals are achieved, but organizational goals are missed

- **Collaboration** is the process whereby a person or persons works with another to produce something. From a business perspective, collaboration is a practice where individuals work together to achieve a common objective. Collaboration allows information sharing, which enables feedback and leads to higher service quality and more creative improvements

Both cooperation and collaboration can be effective and valuable approaches to teamwork and to service relationships

Listening modes

Listening is crucial for efficient communication, and can be improved through practice and training. Poor listening causes mistakes, misunderstandings, and cooperation failures

Different modes of listening can be used in different situations to improve communication. A common scale of listening is proposed by S. Covey in his *7 habits of highly effective people:*

- **Ignoring** - making no effort to listen
- **Pretend listening** - giving the appearance you are listening
- **Selective listening** - only hearing the parts of the conversation that interest you
- **Attentive listening** - paying attention and focusing on what the speaker says
- **Empathetic listening** - listening and responding to understand the speaker's words, intent, and feelings

Diversity

Today's world is very diverse. Therefore, many factors should be considered when cooperating in service relationships, including:

- cultural differences and language
- time zones and locations
- seasonal factors (for example, summer vacations)
- organizational culture

It is always important to:

- foster a respectful attitude
- use the right language
- create an environment where it is safe to express all perspectives
- finish with actionable conclusions
- continually align as the work proceeds

Step 3: Offer

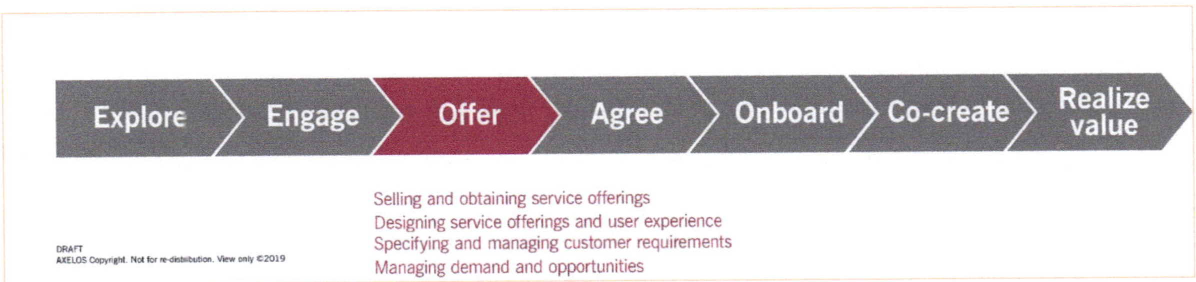

Having obtained a closer service relationship, the customer and the service provider can further shape customer demand and service offerings. This step helps the customer to articulate its needs and demand and the service provider to design matching service offerings

Designing service offerings and UX

Modern methods for software development have enabled a new generation of digital services. These methods have **one important factor in common; they involve the customers and users throughout the design and implementation process of the products and services**

Value-driven and data-driven service design implies an iterative approach based on frequent feedback, continual experimentation, and learning to ensure value co-creation in each step of the design process. This requires engagement, involvement, and interaction between different roles from the service provider and the service consumer

The service provider will bring their expertise on the development methods, but their success will depend on the customer engagement

Lean thinking

Lean thinking can be described as a process improvement philosophy that prioritizes flow efficiency over resource efficiency. In Lean, flow refers to the way work progresses through a system. A work unit can be defined as a piece of work flowing through the value stream. A good flow means the work unit moves steadily and predictably, whereas a bad flow describes a system with a lot of queues where the work unit will have to stop and wait. Lean defines five basic principles:

Lean principle	Explanation
Identify customer value	• The first thing is to uncderstand what the customer needs. **What** creates value for the customer? What is the desired outcome? **Why** is it needed? **When** and **where** is it needed? **How** much? **How** frequent?
Map the value stream	• Next is understanding and mapping the value stream. From the moment a service provider receives a request from the customer for a new or changed product or service, the required activities need to be designed, built, transitioned, and delivered. The key is to define the work unit (request, product, service) and map how it flows through the value chain. Each flow represents a value stream.
Create flow	• The work unit may be subject to several bottlenecks in the value stream. From the perspective of the work unit, This is waste. The flow is improved by eliminating waste.
Establish pull	• When flow is created, the next step is to optimize the value stream. This principle ensures that work is not pushed downstream. Limiting batch size and work in progress. Finishing a few work units at a time.
Strive for perfection	• This principle reflects continual improvement

Table 5.11 The five lean principles

Value stream mapping

Value stream mapping is a Lean technique for illustrating and analysing the logic of a value stream: a method of visualizing the flow from demand/opportunity to value, then planning how that flow can be improved

A value stream map gives a graphical overview of the flow of material and information and identifies areas for improvement. It is a good foundation for understanding how activities are connected and value is created

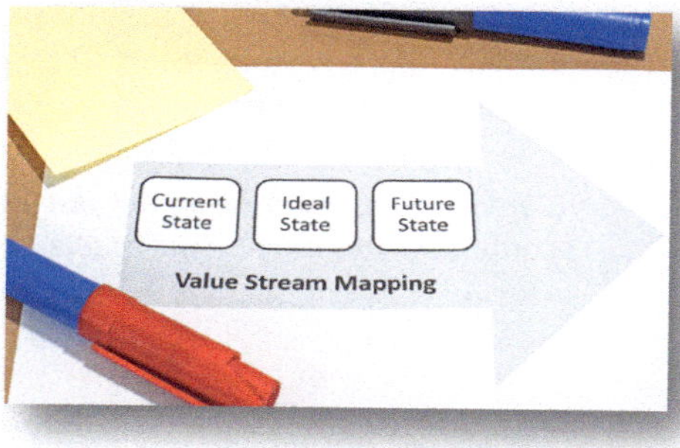

Agile development

One aspect of the Agile philosophy is to apply Lean thinking. Another aspect is iterative development (i.e. SCRUM). Continuous delivery is yet another aspect of the Agile philosophy

Some of the methods used for continuously delivery include:

- **Feature flags -** An "on/off" button programmed into the feature. When the customer is ready, the feature flag can be turned on. If something is wrong, the feature can be turned off again
- **Dark launch -** A new functionality is launched, but the link to the new functionality is not visible before the functionality is mature enough
- **Canary-release -** A dark launch in which a few test users are invited to test the new functionality
- **A/B releases -** A technique used to test which of two alternative solutions is the best. It could be an alternative graphical design or functionality. The consumers are randomly split into two groups and test two different versions of the same feature

These techniques make it possible and safe to develop and release new functionality quickly

User-centred design

User-centred design is an iterative design process which holds the user at the centre of all design decisions throughout the project process

User-centred design ensures that the product and the service focus on what users need and the user experience

It's about balancing user experience with the technical and business requirements

Service-design thinking

Service-design thinking is an effective way of tackling the problem of problem-solving

It encourages users to define value and is a method that continuously gathers feedback on what is and is not working

The ultimate objective of the service-design thinking process is to identify solutions to the original challenge that are desirable, feasible, and viable

Service blueprinting

A blueprint is a diagram visualizing the service usage that aims to optimize the user experience. In a service blueprint, the key elements are separated with three horizontal lines:

1. **The line of interaction**
   - This pinpoints the direct service interactions between the customer/user and the service provider
2. **The line of visibility**
   - This separates the service activities that are visible to the customer and users from those that are not visible. Visible activities appear above this line and invisible activities appear below this line
3. **The line of internal interaction**
   - This separates the employees in contact from those who do not directly support interactions with customers and users

For one service, there may be multiple blueprints if there are several different scenarios. Drafting a blueprint, it is important to proceed from the top line and work down

Service blueprint (example)

Service blueprints are useful in order to:

- link the customer journey to products and services
- highlight user touchpoints and service interactions
- discover weaknesses and identify opportunities for optimization
- bridge cross-department efforts and avoid double workload

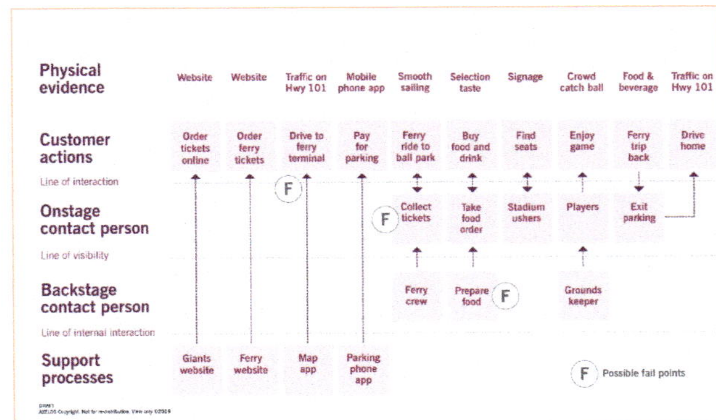

Figure 5.7 Example service blueprint.

Designing for onboarding

As part of a user-centred approach, the onboarding approach should be defined for every product, service, and service offering as part of their design to smooth onboarding

A documented onboarding approach is used for planning an onboarding initiative

Approaches include scope, actions, stakeholders, timelines, and other aspects of onboarding

When defining the onboarding approach and planning onboarding initiatives, organizations should consider all four dimensions of service management. It should also be influenced by the external factors

One way to structure the onboarding approach is to follow the steps of the continual improvement model

Structuring the onboarding approach

Continual improvement steps	Onboarding approach
What is the vision?	Define the expected result of the onboarding
Where are we now?	Establish the baseline and determine the scope of the onboarding, including the required resources
Where do we want to be?	Define the onboarding goals, including the ideal combination and interaction of the resources
How do we get there?	Plan onboarding actions and responsibilities
Take action	Run the onboarding actions according to plan
Did we get there?	Control the onboarding and ensure its success: controls and metrics compared to baseline and goals
How do we keep the momentum going?	Review and improve

Table 5.13 The continual improvement model and the onboarding approach

Key activities for onboarding design

When an onboarding approach is being designed as part of product, service, and service offering, key activities include:

- identifying those provider's resources that will interact with consumer's resources
- identifying those consumer's resources that will interact with provider's resources
- identifying the need for introducing each pair of resources
- exploring opportunities to optimize and automate introductions
- creating procedures where manual or human-controlled introduction is required
- testing the procedures and updating them based on test results (repeat as needed)
- documenting and communicating to the involved parties

Pricing

Several factors should be considered when deciding the price of a service:

- How good is the service provider at communicating the value of the service?
- How much are the customers willing to pay?
- Are there similar services in the market?
- How are similar services priced?
- Is it easy to scale the service?
- Is it easy to onboard new customers?
- Could the service be used as a loss leader?
- How mature is the service provider?
- Can the three bottom lines approach influence the pricing?

In some cases, pricing can be used in the short term to undermine competition

Pricing options

Pricing option	Description	When applicable
Cost	• This option is based on a break-even or cost recovery model. The chargeable item is priced as close as possible to the actual cost of the cost units	• For internal service providers or non- profit organizations
Cost plus	• The mark-up (x%) can either be set by the service provider to match returns on other investments or allocated by the service provider to meet strategic business needs	• The mark-up can be used for encouraging the use of new products and services but discouraging the use of legacy products and services
Market price/Going rate	• The price is comparable with similar service offerings on the market	• Commodity and out-of-the –box services. It is important to remember that sourcing services externally offsets some of the commercial risk from the customer to the service supplier
Fixed price	• The service provider sets a price based on negotiations with the customer, which covers a set period and a predicted consumption	• Enables the parties to lock the price regardless of cost fluctuations, for example, if a customer has a fixed budget
Differential charging	• Setting different charges for different usage of the same or similar services at different times	• Enables an organization to reward some usage patterns over others, for example, discouraging service usage during peak time periods

Table 5.14 Pricing options

Internal sales

Some benefits of selling to internal customers are:

- enhanced utilization of existing services
- better control of the service demand
- improved communication with customers and users
- feedback on how well services meets the needs

Internal sales contribute to the overall customer perception of the internal service provider

One of the most important tools for the internal sales process is **the service catalogue**: a communication tool for promoting services. For the service provider, it helps define and make products and services visible. The service catalogue is vital for relationship management, promoting existing services, and for discovering the need for new services. It is also a key tool for service level management when defining services levels

External sales

External customer sales is more like traditional sale techniques, such as advertising and sale campaigns

The sales process depends on the type of service relationship, and contextual factors. These include:

- In highly-regulated environments and in the public sector, acquiring goods and services may be subject to regulation; certain service providers may have been preauthorized to sell their products and services
- Some service consumers require that purchasing is done by a procurement team
- Others may have defined formalized purchasing processes that include formal request methods, such as requests for information, requests for quotation, requests for proposal, proof of concept, demos, and so on

Methods for requesting products and services

Request for information	Request for quotation	Request for proposal
• Gathering information to identify potential vendors • When you are looking for information • Ask general questions • Causal • Fast	• Targeted request for pricing for a highly specific product or service • When you know exactly what you need • Ask questions about requirements • Structured • Focus on price	• Strategic and intensive proposal process • When you want to compare proposals • Ask specific questions • Formal • Comparison on vendors

Table 5.15 Different methods for requesting products and services

Step 5: Onboard

Onboarding includes all the necessary activities that enable a service consumer to start using the service and a service provider to be ready to deliver the service

These could range from the service being turned on and made available to use (for example, a mobile phone connecting to a network) to contractual agreements, user awareness, training, and sharing of resources (for example, outsourcing desktop provision)

Onboarding

Effective onboarding enables service provision and consumption, increases service usage efficiency, improves user experience, ensures satisfaction, and enriches the relationship between the service provider and the service consumer

Onboarding occurs after an agreement is reached or changed but before service consumption starts. Onboarding creates the first service impression for users, which may strongly impact any further decisions about the service relationship made by sponsors and customers. Each onboarding initiative should be carefully planned and managed in line with the agreed plans

Onboarding includes:

- building awareness about the new service consumers (or the new scope of a service relationship with existing consumers) among stakeholders
- ensuring that all resources within the scope of the service offering are prepared for the service provision
- ensuring that customers and users are ready for the service consumption

The purpose of onboarding and offboarding

The table summarizes why the service provider, service consumer, and other stakeholders should invest in effective onboarding and offboarding

Onboard	For the service consumer	For the service provider
Facilitate outcome and experience	• To ensure a better return on investments through effective use of services • To improve user experience • To increase the effectiveness and efficiency of business operations through the effective use of services • To maximize value from cooperation with new service providers	• To maximize value from cooperation with new service consumers/customers/users • To improve perception of the new services and the service provider in general • To increase customer and user loyalty and engagement
Optimize risk and compliance	• To decrease the probability of user incidents and questions related to new services and users • To decrease the duration of • transition times to new services/providers	• To decrease the probability of incidents and associated breaches in service quality • To prevent/decrease user • resistance to new services and/or service providers
Optimize resources and minimize cost	• To reduce the costs and losses associated with the transition to new services/providers • To optimize the cost of user training • To optimize the costs of user support	• To reduce transition costs To reduce user support costs • To optimize onboarding costs and overall resource utilization

Table 7.1 The purposes of onboarding and offboarding

Planning on onboarding

Planning an onboarding initiative is a collaboration between the customer and the provider and the customer is involved in:

- defining the onboarding goals and associated metrics
- identifying the service provider and service consumer resources that are
- covered by the onboarding (access and integration)
- planning onboarding actions, including schedule and responsibilities
- reviewing and accepting the onboarding plan

An **onboarding plan** should answer the following questions:

- What are the goals of the onboarding?
- What is the scope of the onboarding?
- What are the onboarding actions?
- Who is responsible for the onboarding actions?
- How can the onboarding be controlled and its success ensured?

Onboarding goals

Service providers should define, agree, and build awareness about the onboarding goals with relevant stakeholders. The onboarding goals should be defined in the context of every onboarding initiative

Examples of the onboarding goals are:

- to ensure a smooth migration of the service consumer to the agreed services
- to ensure a smooth migration of the service consumer from an in-house technology platform to the cloud
- to support the temporary increase in the number of users of the selected services
- to support the service consumer's switch from one (third party) supplier to another

Onboarding success should be evaluated on the basis of the agreed goals **(outcomes),** rather than just checking the progress and completion of the planned onboarding actions **(outputs)**

Onboarding scope

To define onboarding's scope, the following questions should be considered:

- What are the consumer resources requiring onboarding?
- What are the provider resources required for onboarding?
- When does onboarding start and end?

All these questions are expected to be answered by the onboarding approach, but every onboarding initiative requires the review and adaption of the approach, depending on the **scope of the initiative**

The onboarding approach will affect one or more resources. When creating an onboarding plan for an onboarding initiative, the service provider should identify specific resources that require onboarding and required actions based on the onboarding approach

Onboarding actions

After agreeing the scope of the onboarding initiative, onboarding actions should be planned. Methods and tools used for the planning may differ based on the scope and scale of the onboarding initiative and on the organization's policies, rules, and practices

In many cases, especially when services are provided to individual consumers, service providers standardize and automate onboarding. This is achieved by using highly standardized and compatible IT solutions, such as common platforms, standard libraries and protocols, microservices, and so on. Seamless onboarding is usually maintained by the following factors:

- multi-language interfaces
- highly-intuitive interfaces
- highly-available support, including self-service and peer support
- automated software installation and updates
- cross-platform availability

Onboarding control

When planning an onboarding initiative, it is important to agree on the approach to the control and validation techniques of the initiative success

A formal review of the completed onboarding initiative may be a valuable exercise

The review may include the following:

- formal confirmation that all planned onboarding actions have been completed
- review of users', customers', and other stakeholders' satisfaction and experience
- review of outstanding actions and errors
- risk assessment
- registration of improvements for the continual improvement practice

Onboarding control

The onboarding review may lead to various improvements to:

- product, service, and service offering design
- user interfaces and procedures
- service portfolio
- service catalogues
- relationships with partners and suppliers
- service provider's management practices
- ongoing onboarding plans and initiatives

Relating with users and fostering relationships

When users interact with a service provider, **the user experience becomes a critical contribution to the success of the service**. Poor user experience leads to poor productivity and creates a negative perception towards the service and the service provider

This means that the service provider and the customer should be conscious of user experience at all steps of the customer journey. A positive user experience should be one of the most important requirements for any service that includes direct user interactions

User experience should be considered a part of product and service design, development, testing, transition to the live environment, ongoing delivery, and regular assessment and review. It should also be a subject for continual improvement

User onboarding influences user attitude towards the service and the service provider. Therefore, it is important for successful service consumption because it enables value co-creation and helps to establish and maintain a sustainable relationship with users and the consumer organization

Relationships with corporate users

To **prevent resistance** to new services and enable good relationships, service consumer and service provider organizations should work together in every step of the customer journey by:

- considering how user experience is impacted at each step of the customer journey
- planning user onboarding as part of every new service implementation
- practicing organizational change management
- involving users in requirements articulation
- involving users in testing the services and onboarding actions
- designing user-friendly interfaces
- understanding and leveraging people's experiences as private users and their related expectations
- providing a useful, convenient, and relevant service catalogue, including a service request catalogue
- monitoring user satisfaction constantly to improve service experience
- involving users in service testing, assessments, and reviews
- involving influential users in service promotion and peer-to-peer support programmes
- fostering user communities and actively supporting their members
- performing service usage analytics for the profiling and proactive use of real-time end-user computing data

Relationship management with individual consumers

Step	Challenges
Explore	Service consumers do not explicitly express their needs and expectations Sometimes service consumers are unaware of their needs and opportunities Individual opinions do not necessarily represent the needs of a larger consumer group
Engage	Individual face-to-face contacts are often impossible or inefficient due to a large number of users and limited capacity of service providers Pre-sale and sales are highly regulated to protect consumer rights Communities', influencers', and peers' opinions are important for an initial engagement
Offer ***Agree***	Large number of service consumers Need for simple and fast contracting procedures and interfaces High level of international, national, and industry regulation Need for simple and unambiguous language
Onboard	Large number of service consumers with different skills and backgrounds Need for simple and fast onboarding Different technology background
Co-create	Large number of service consumers with different skills and backgrounds Different technology backgrounds Different language and communication skills High exposure of user experience and opinions via social media
Realize	Large number of service consumers Low impact of individual experience on service quality and satisfaction statistics Inefficiency or impossibility of direct service review with every service consumer Different expectations and needs supported by the services

Table 7.6 Relationship management with individual service consumers

Delivery channels

It is important to establish the appropriate user engagement and delivery channels in order to provide a good user experience. The purpose of omnichannel management is to provide a seamless user journey to the user. These principles are illustrated in the figure below.

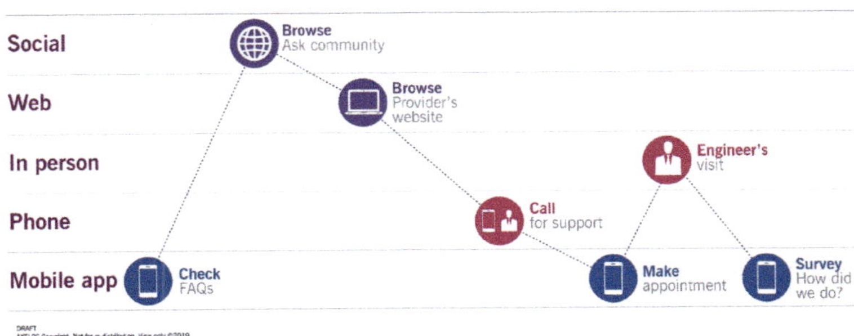

Figure 7.2 Seamless user journey with omnichannel management

True omnichannel support that focuses on the user is only obtained when different approaches are orchestrated into a seamless user support experience. This can be done by:

- uniquely identifying and recognizing users across all channels
- systematically collecting and analysing user data
- leveraging user data across all encounters
- monitoring and managing performance across all user journeys

Enabling users for service

The service catalogue management and service desk practices are particularly important for user onboarding. These practices ensure effective and friendly interfaces for users during the steps of the user journey

To effectively enable and contribute to offering user services (including user awareness of new services available and the associated service requests), user-facing service catalogue should:

- be structured in a logical way, reflecting users' needs and patterns of activity
- be presented in a language that is clear and familiar to the users
- only include services that are relevant for the users (and that are available or already being provided)
- include services and associated service requests
- be up to date
- be actionable (where possible, automated for actions users are eligible to do, such as minor changes to the level of services and initiating service requests

Optimizing user capabilities

Service relationships are about value co-creation for all stakeholders. Each service interaction is an opportunity to elevate the capabilities of another party. Examples of service providers and customers using the ITIL guiding principles to improve user capabilities:

Guiding principle	Example of application to optimize user capabilities
Focus on value	Users should understand the purpose and context of their work and of service usage. They should be encouraged to offer service improvements that may contribute to value co-creation
Start where you are	The improvement of user experience should be based on current practices, habits, and expectations. Radical changes in user experiences are rarely perceived as changes for the better and are often resisted by users
Progress iteratively with feedback	All changes in requirements to users, service delivery and assessment, service usage procedures, and other aspects of user experience should be tested and continually reviewed based on user feedback. Users should be encouraged to provide their feedback and feedback follow-up actions should be transparent to the user community
Collaborate and promote visibility	Where joint operations are required, users should understand the need for collaboration and help each other, as well as service providers, partners and suppliers, and other parties involved. When services do not operate as intended, or when a user does not know how to use it, calling for help or reporting an incident, should be safe, easy, and encouraged
Think and work holistically	Services and their role in the value co-creation should be visible and transparent to all involved parties. Users should understand the context of their work and dependencies
Keep it simple and practical	User interfaces and all other touchpoints should be as simple as possible. Users should have an interface to propose improvements to the interface, and these proposals should be treated seriously and transparently
Optimize and automate	Continual optimization and automation of user experience should be a theme of all touchpoints and service interactions between users and service providers

Table 7.8 Examples of service providers and customers using the ITIL guiding principles to improve user capabilities

Offboarding customers and users

Just as with onboarding, offboarding actions and responsibilities should be predefined as part of product and service design and then adapted for the context of a specific onboarding/offboarding initiative

When a service consumer leaves as a result of dissatisfaction or a dispute, it is important to identify the reasons. In these cases, it is even more important that offboarding has been agreed and prepared properly to ensure a seamless offboarding, where each step of the process is agreed in a conflict free environment

Additionally, both parties should be careful not to escalate any conflicts or cause harm for any other party; they should agree to keep disputes confidential. This is especially important when subsequent legal action is expected

Customer offboarding

Is performed by the service provider upon agreement expiration or termination. Actions include:

- communicating about the planned service termination to all relevant stakeholders, including users, customers, relevant partners/suppliers, and regulators
- responding to any users requesting further information or other support
- organizing and performing equipment hand-off from the service consumer to the service provider
- removing any service provider resources that have been operating on the service consumer's premises
- revoking access of either party to the other party's resources, wherever applicable
- archiving and retaining records
- calculating and processing exit payments, including outstanding amounts on either side
- changing third party contracts associated with the services being terminated, such as supporting technical services, insurance, and others
- maintaining formal offboarding records as agreed by the parties and/or required by applicable regulations
- performing relationship management actions relevant to the situation, such as closing meetings, letters of appreciation, invitations to restore service relationship, and more

A more complex modern approach to offboarding is the co-operation between service providers that includes mutual offboarding agreements. In these cases, new providers represent the service consumer to the old provider. Switching a service provider may also involve third parties

User offboarding

User offboarding may occur as part of an ongoing service relationship, without service or contract termination. User offboarding usually includes:

- communicating about the planned offboarding and associated responsibilities of the user
- interacting with users requesting further information or any other support with regard to the planned offboarding
- organizing equipment hand-off from the user to the service provider, or to a responsible representative of the service consumer
- changing or cancelling the user's access
- securing records by archiving and retaining
- deleting information that is not archived
- maintaining the formal offboarding records agreed by the parties and/or required by applicable regulations
- performing relationship management actions, such as closing meetings, composing letters of appreciation, and so on

Offboarding - Summary

To proceed from agreement to service provision and consumption, the parties must undergo a transition that involves the integration or separation of the resources of the service provider and the service consumer.

This approach should be defined as part of service design and onboarding or offboarding activities should be planned, run, and controlled accordingly.

The main onboarding activities include fostering user relationships, orchestrating omnichannel access, enabling users for service, and elevating mutual capabilities

Step 6: Co-create

When all parties are ready for service delivery and consumption, the next step of the journey begins. It is arguably the most important; all previous steps aim to enable it and ensure that it will be successful

The purpose of this step is for those involved in the service relationship to **act together to ensure continual value co-creation** based on agreed service offerings

The scope of the provision and consumption step includes:

- service delivery and support
- service consumption
- service usage

Ongoing service interactions

There are different approaches to provide services to users. The most common options include:

- **Pull or push** Service actions may be initiated by users, for example pull. That is the case when the user requests a service on a service portal, downloads an application, calls a service desk and so on. They may also be initiated by the service provider, such as push. For example, the user may get a report every month as part of a subscription or the service provider may update the software on the user's device automatically when a new version is available.
- **Automated or manual** Initiation and implementation of service actions can be automated in different ways (*see next slide*)
- **Tailored or out-of-the-box** Interactions between users and service providers can be standardized for all or for large groups of users or adjusted to fit the requirements and preferences of individual users or small groups
- **Direct or indirect** Service providers can interact with users using their own resources and employees, or through agents. For example, an airport agent is often used to check in the traveller on behalf of the airline. The service consumer can also provide the service on behalf of the service provider. For example, the customer has agreed a service with an outsourcing company, but the users are still served by the service desk of the service consumer

Service technology encounters

Automated or manual Initiation and implementation of service actions can be automated in different ways:

- **Technology-free.** Technology is not involved in the service provision
- **Technology-assisted.** Technology is used to support the service provider in the service provision
- **Technology-facilitated.** Both the service provider and the user have access to the same technology as part of the service provision
- **Technology-mediated.** Service is provisioned without the service consumer and the user being in physical proximity
- **Technology-generated.** The service provider, the service consumer, or both are represented entirely by technology

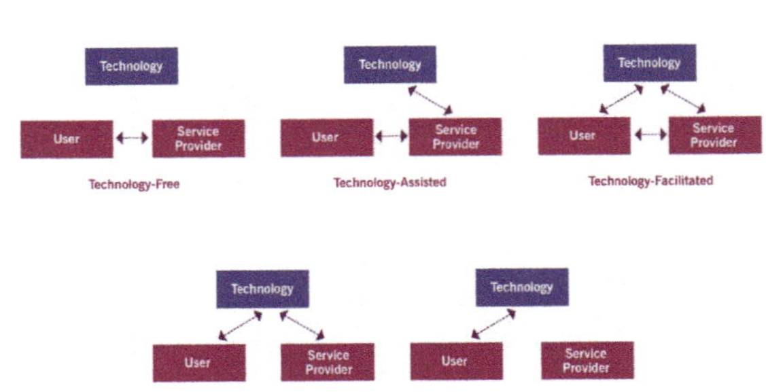

Figure 8.2 Types of service technology encounters

Service requests

For all types of service requests, rules and conditions should be agreed and communicated to all of the involved roles in both provider and consumer organizations. Service request management contributes to the success of service usage, this provision may be enhanced by:

- user-friendly, interactive service request catalogues, tailored for user's status and level access request
- self-service interfaces for initiating and fulfilling service requests
- status updates and push notifications about ongoing requests
- proactive communications about service status, requests availability and new options
- friendly and professional support agents fulfilling requests
- automated fulfilment and delivery, wherever possible and preferred by users
- collecting and processing feedback
- effective workflows for request fulfilment
- effective integration with other management practices

Definition: Service request
A request from a user or a user's authorized representative that initiates a service action which has been agreed as a normal part of service delivery

Service desk interaction

It is important for the service relationship to be successful and that service desk interfaces, procedures, and rules are agreed by the parties and communicated to all users at the onboarding step

These interfaces and procedures should be simple, user-friendly, and easy to follow. The service desk practice should be continually improved based on feedback from users to ensure a positive user experience

Key message:
The purpose of the service desk practice is to capture demand for incident resolution and service requests. It should also be the entry point and single point of contact for the service provider with all of its users

User queries triage criteria and key practices

Triaging criteria	Service query types	Key ITIL practices
Agreed SLAs are breached; or user reports a negative experience	Incidents	Incident management
Agreed conditions for an emergency situation are met; or the query can be linked to an ongoing registered disaster	Disasters	Service continuity management
Valid service request option is selected from the service request catalogue	Service requests	Service request management
Valid change option is selected from service catalogue which is available to the user	Change requests	Change enablement
User reports negative experience that is known as not breaching any agreed SLA (complaint); or user reports a positive experience for the record	Complaints and compliments	Relationship management
User proposes improvements that are not covered by available options of service catalogue (incl. service request catalogue)	Improvement proposals	Continual improvement
User reports suspicious activities or events that may indicate threats to information security	Suspicious activities and events	Information security management

Table 8.4 User queries triage criteria and key practices involved in their processing

The concept of "Moment of truth"

A moment of truth is a key touchpoint or service interaction between the service provider and a user in which the user forms or changes his or her impression of any aspect of the service experience, service organization, products, or services

Therefore, the service provider should be aware of potential moments of truth. The service provider organization should possess the resources and capabilities to make the right decisions and provide the right service

Essentially, the service provider should consider aiming to turn negative events into positive moments of truth

Intelligent disobedience

Intelligent disobedience is the act of breaking the rules to do the right thing. In terms of service relationships, perception advantages typically include factors such as familiarity, direct involvement with a service, or awareness of a user or customer's needs

Intelligent disobedience is simple and common in our everyday lives and, in most cases, it can be helpful and genuine. Minor displays of intelligent disobedience that improve our lives are regularly appreciated; for example, a shop manager who delays closing to allow customers to finish their shopping

There will always be unexpected issues to manage in which existing rules are not helpful or sensible. In these cases, the service experience depends on the staff who interact with customers and users correctly assessing whether the rules should be followed or suspended. This means that, in order to maximize the value delivered to customers, providers should:

- recognize that the rules in place will not always be appropriate to the situation
- encourage and equip staff with the knowledge and skills to identify legitimate customer requirements and to perceive and fulfil those requirements
- create an environment where staff feel confident to act outside the rules when dictated to do so by the circumstances
- ensure that actions are documented to improve rules and future understanding

Customer and user feedback

Customer and user feedback is important for the success of the service relationship and continual improvement of services

Service providers are interested in open, direct, and comprehensive opinions about their services and related user and customer experiences. So there is an important task for service providers is gathering and processing effective feedback

However, many users and customers are reluctant to provide open and honest feedback. This reluctance is often related to past negative experiences

Nurturing user communities

User communities can be valuable components of service interactions during normal operations, incidents, and other difficulties. Peer support, knowledge articles, discussions, and improvement initiatives can help to decrease users' demand for support from the service provider and move support closer to the user (shift left), provide valuable insights and improvement opportunities, and help in communicating news and status updates

In many cases, user communities form naturally, especially when services are provided to many individual and corporate service consumers. However, they can also emerge within organizations, particularly if users do not feel enough support from the service provider. Either way, communities can become a powerful means of communication and cooperation between users and service providers

User communities can be established by the service provider and unite all service users. They also can be formed by user groups and cover multiple services from multiple service providers

Step 7: Realize

Realizing value for the service provider
Evaluating value realization and improving customer journey
Assessing and reporting value realization
Tracking value realization
Realizing value in different settings

The purpose of this step is to track, assess, and evaluate whether the value that the stakeholders expected and planned for is realized throughout all steps of the journey and to continually identify improvements to the journey and the services

ITIL planning and evaluation model

- A requirement for assessing and evaluating value realization is **tracking and measuring service value indicators**.
- A service value indicator is a measure that either directly or indirectly indicates the situation or level of a specific aspect of service value. Indicators reflect the achievement of an objective. In this case, the objective is the desired or agreed service value.
- Indicators are reinforced by one or more metrics. A metric is an important characteristic of value that can be expressed in quantifiable units based on data collected by a procedure or technical means. Metrics can be defined for customer journeys, services, products, practices, and resources of all four dimensions of service management.
- The hierarchy from purpose to metrics is described in ITIL planning and evaluation model, which is shown in the figure to the right.
- Tracking value realization includes the following activities:
    - identifying the direct and indirect indicators of service value and the links between them
    - defining and measuring the underpinning metrics
    - capturing measurement data.

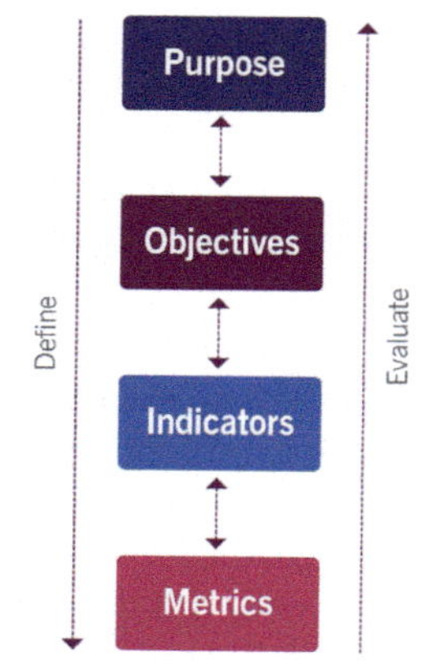

Figure 9.2 ITIL planning and evaluation model

Tracking performance, output and outcome

It is often harder to identify direct outcome indicators than indirect outcome indicators, such as output and performance indicators. If clear links can be established between outcome indicators and output and performance indicators, the latter can be used to indirectly track service outcome

One way of linking performance and output indicators to outcome is to use the service stack model as a basis, considering the lower layers as value drivers for the higher layers. The service stack is shown again in the figure to the right

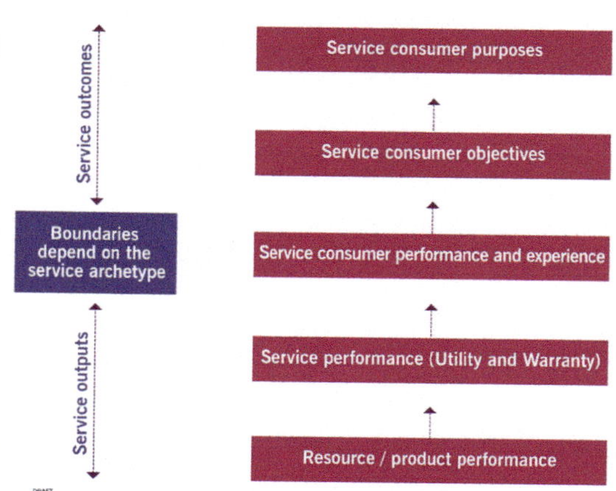

Figure 9.3 Levels of value realization tracking

The service profit chain

Another useful model for linking performance and output indicators to service outcome is the service profit chain proposed by Heskett

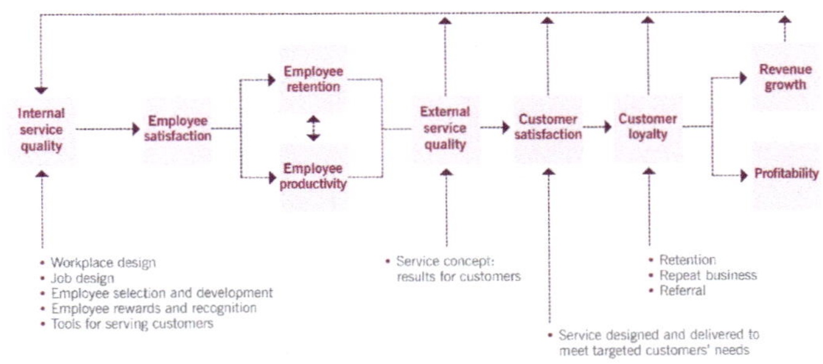

Figure 9.4 The service profit chain

According to the service profit chain model, profit and growth are stimulated primarily by customer loyalty. Loyalty is a direct result of customer satisfaction, which is largely influenced by value of services provided to customers

Value is created by satisfied, loyal, and productive employees. Employee satisfaction results from high-quality support services and policies that enable employees to deliver a good service

Tracking experience and satisfaction

In order to ensure that the customer and users are satisfied with the service, and the whole customer journey, the service provider should measure and track customer experience, perform customer satisfaction surveys, and request and process feedback from service consumers.

Customer experience and user experience are the sums of functional and emotional interactions with a service and service provider as perceived by a customer or user. In order to completely understand customer experience and user experience, there are three questions to consider from the customer perspective:

- How does the service work? (functional experience)
- How does the service feel? (emotional experience)
- To what degree does the service fulfil my needs? (satisfaction)

Tracking service usage

Service metering and service usage analytics is a way to track, map, and understand customer and user behaviour. This contributes to the understanding of how and when value is realized

Services are designed to enable customer activities, which in turn enables the achievement of customer outcomes. Every time a customer activity is performed, it generates demand for services

Customer assets, such as people, processes, and applications all perform activities. Because of the way these assets are organized or the tasks they are completing, this activity will tend to be performed in patterns. These **patterns of business activity (PBA)** represent the dynamics of the business and include service interactions with customers, suppliers, partners, and other stakeholders

Assessing value realization

Assessing and reporting value realization is about consolidating data from various data sources, correlating the consolidated data, interpreting and assessing the information, and presenting it in a form that is appropriate for making decisions. It may include formal service reviews or informal meetings between the service provider and the customer

Key message:

The customer may seek evidence on service outcomes achievement from the service provider at different levels, such as return on investment, achieving strategic objectives, process performance, satisfaction measures, service levels, and more

However, as we have seen, these levels are all connected. Even in cases where the customer only asks for service availability, capacity, and performance:

- The service provider should try to map service performance metrics to consumer process performance indicators, strategic objectives, and consumer purpose in order to really understand the customer
- The customer should link service level with its strategic objectives and, at least qualitatively, analyse how well the service contributes to a service consumer purpose

Assessing value realization

As part of the assessment, the captured experience, performance, and output data should be correlated with outcomes, risks, and costs and the overall service contribution to the customer objectives and purposes should be assessed. Once data is consolidated, the results should be analysed and the following questions should be considered:

- Are the targets being met?
- What are the dynamics?
- Are there any clear trends?
- Are the trends positive or negative?
- Are there underlying structural problems?
- Are improvements required?
- Are there unforeseen patterns or unexpected and unplanned outcomes that should be fed back to design and improvement?
- Is there new information about the products, services, customer journey, or context (market)?

If applicable, assessment techniques, such as post-implementation reviews, audits, and cost benefit analyses should be used

Evaluating value realization

In the 1970s, Chris Argyris and Donald Schon introduced the concept of single- and double-loop learning as a part of their organizational learning theory

- **Single-loop learning** describes the type of learning that takes place when the objective is to fix problems within the present organizational structure so that the system will function better. It does not attempt to alter the structure of the system

- **Double-loop learning**, also known as reframing, questions the purpose and function of work being done within an organization and does not take existing organizational structures for granted. Double-loop learning is concerned with understanding the basis for the tasks being completed, rather than a more efficient process for completing them

Evaluating value realization

Evaluating value realization not only verifies that individual services create the desired value and that the completed end-to-end journeys leave the desired customer and user experience behind (single-loop learning), but also that the involved SVSs are fit for purpose (double-loop learning), as is shown in the table below

	Single loop learning	Double loop learning
The services	• Do services lead to the co-creation of the desired value?	• Are the SVSs fit for purpose?
The customer journey	• Do end-to-end journeys leave the desired customer and user experience behind?	• Are the SVS fit for purpose?

Table 9.6 Evaluating and improving services and customer journeys

Evaluation and verification

Evaluation is a chance to challenge initial assumptions. To solve a customer's problem, the following questions should be considered in order to re-evaluate the assumptions:

- Is there still a real problem that needs to be solved?
- Is a service still the best way to solve the problem?
- Is the service still fit for purpose and fit for use?

As conditions always change, the customer should keep the service provider informed about changes that may affect service outcomes, experience, risks, and costs.

Continual improvement

Evaluation of value realization provides vital input for the continual improvement of products, services, and all components of the SVS. Other sources for improvement are:

- service usage analytics
- incident, complaint, and problem analysis
- analysis of service request patterns
- analysis of self-service patterns and usage of knowledge articles
- change requests and improvement requests
- user feedback and feedback from user communities
- customer feedback and customer satisfaction surveys
- changes in service demand

These sources provide information about ongoing service provision and consumption enabling capacity and availability predictions, exploration of new opportunities, identification of new service value streams, etc.

Assessing outcomes

First and foremost, the service provider exists to facilitate value for the customers and users, not to pursue its own business goals. Tracking, assessing, & evaluating outcomes should there-fore ideally be incorporated in the tracking, assessment, and reporting of customer outcomes

On the other hand, an external service provider pursues its own business goals, although it serves its customers and helps them achieve their goals as a means to achieve its own. These goals are not only financial but may also include brand recognition, market share, capability elevation, resource utilization, and more

Therefore, the service provider not only needs to track customer value realization but also its own value realization

Charging and billing

Part of the service provider's value realization is profit, or at least cost recovery. The amount customers will be charged for a service depends on the charging policies. Several options exist for this decision:

- **Cost recovery or break-even** In this case, the service provider will only seek to recover its costs. It will not make a profit or loss
- **Recovery with an additional margin** In this case, the service provider will seek to recover more than its actual costs. It is important to note that the margin is not a profit. The additional funding may be placed in a separate account, set aside for refreshing technology or covering the costs of unanticipated projects, or it may be used to offset the costs of new investments
- **Cross-subsidization** In this case, a subset of services is charged with an additional margin, which is then applied to offset the cost of another subset of services. This policy is often used to fund services to internal customers that have run out of budget, or whose business performance is lower than expected. The risk here is that temporary subsidies often become permanent
- **Profit** In this case, the service provider aims at making a profit for its owners or for reinvestment in the business

Closing

We have described the seven steps on a generic customer journey and provided guidance on how they contributes to drive stakeholder value and experience. The figure to the right shows the entire generic customer journey

As we have seen, a customer journey never follows a linear or theoretical sequence of steps in real life. It is more the rule than the exception that a customer journey follows an emerging pattern of touchpoints and service interactions

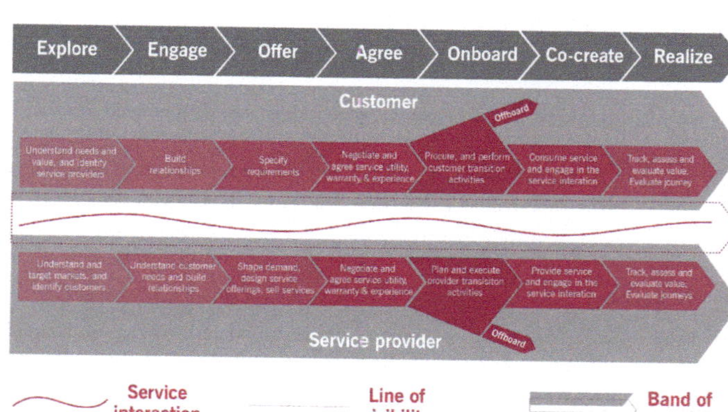

Figure 10.1 The generic customer journey

Service is not manufacturing or production, but value co-creation between two or more stakeholders. The service experience is formed by individual touchpoints and service interactions on the customer journey. To improve service consumer satisfaction, each touchpoint must lead to good customer experience so that the whole journey delivers on customer expectations

Summary

We have just talked about:

- ✓ The concept of customer journey
- ✓ The concepts mutual readiness and maturity
- ✓ How to foster stakeholder relationship
- ✓ How to onboard and offboard customers and users
- ✓ How to act together to ensure continual value co-creation
- ✓ How to realise and validate service value

Q: A service provider is launching a new service. The target market is users who have limited experience of using the internet and are unlikely to use social media. Which is the best method of providing user support?

A. Provide simple on-line support and provide contact numbers for the service desk

B. Use machine learning chatbots to anticipate the needs of the users and provide solutions

C. Implement a 'shift left' approach to provide support and downloadable help articles

D. Use popular networking sites to promote and provide on-line user support

Q: Which describes user-centred service design?

A. Building a prototype on the minimum functionality that can be produced quickly

B. Applying the MoSCoW technique to a set of user requirements

C. Balancing user experience with the technical and business requirements

D. Using value stream mapping to identify a set of user requirements

Q: What is an attribute of a strategic partnership?

A. Minimal information is shared

B. A high level of trust needs to be developed

C. The customer needs to be able to exit easily

D. The relationship is driven by price

Q: Which statement about the end-to-end customer journey is CORRECT?

A. It focuses on achieving outcomes

B. It represents a pre-determined path

C. It reflects an overall perception

D. It involves optimizing touchpoints

Q: A service provider encourages their users to form a user support community. What is the potential benefit for the service provider?

A. Reduction in demand for support from the service provider

B. Creation of groups that can be used to provision services

C. Improved collaboration within the service provider organisation

D. Reduced collaboration between user organisations

Course schedule

- ✓ Day 1: ITIL 4 Foundation
  - Key concepts, 4 dimensions of SM, ITIL Service Value System, the service value chain & 7 guiding principles
- ✓ Day 2: Create, Deliver and Support
  - How to use Value streams with support of ITIL practices
- ✓ Day 3: Drive Stakeholder Value
  - Customer journeys, relationships and service offerings
  - How to act together to ensure continual value co-creation
- Day 4: High Velocity IT
  - The Digital enterprise and how to use guiding principles
- Day 5: Deliver, Plan and Improve and Exam
  - Direction and Planning, GRC, Communication & OCM
  - Work with sample exam and examination

ITIL 4:
HIGH VELOCITY IT

Key learning requirements

HIGH VELOCITY IT

- Understand concepts regarding the high-velocity nature of the digital enterprise, including the demand it places on IT
- Understand the digital product lifecycle in terms of the ITIL 'operating model'
- Understand the importance of the ITIL Guiding Principles and other fundamental concepts for delivering high velocity IT

Introduction

The scope of the High Velocity IT (HVIT) book is the primary activities in the digital value chain

In other words, what the practitioner does and which resources they use across the lifecycle of digital products, in order to:

- Make the right digital investments
- Realize and deliver digital products and services quickly
- Provide digital products and services that are highly resilient to disruption
- Ensure that the service consumer realizes value from the digital products and services
- Assure conformance of activities with governance, risk and compliance requirements

Although primary and supporting activities cannot be treated in isolation from each other, the book does not describe supporting activities such as how resources are acquired or developed, or how the resources and activities are managed and improved

Key learning requirements

Terms:

- Digital organisation
- High velocity IT
- Digital transformation
- IT transformation
- Digital products
- Digital technology

Five objectives (to achieve with digital):

- Valuable investments
- Fast development
- Resilient operations
- Co-created value
- Assured conformance

The Digital Product Lifecycle:

- Service consumer and service provider have different perspectives on digital products. They each have their own lifecycles, that overlap during the period of engagement

Principles, models and concepts:

- Ethics
- Safety culture
- Lean culture
- Toyota Kata
- Lean / Agile / Resilient / Continuous
- Service-dominant logic
- Design thinking
- Complexity thinking

Digital organisation

Definition: Digital organisation
An organization which is enabled by digital technology to do business significantly differently, or to do significantly different business

'Digitally enabled organisation' might be a better term

Digital technology is:

- A significant enabler for internal processes
- Often part of their products and services
- A strategic part of the business model
- For primary activities, not just supporting

'Digital first' is part of the culture of the organisation

High velocity IT

It's not binary, can be higher or lower

High velocity does not reduce service utility or warranty

Velocity is speed *and* direction

- Must be doing the right thing fast

Definition: High velocity IT
The application of digital technology for significant business enablement, where time to market, time to customer, time to change, and speed in general is crucial. High velocity is not restricted to fast development, and is required from innovation through development and operations to the actual realization of value

Digital transformation & IT transformation

Enables an organisation to do business significantly differently OR to do significantly different business

Transformation means major change

- But could be agile and incremental

Often includes major investment in:

- Digitizing, robotizing, or automation

Often requires organizational change

- Because digital solutions are used differently

Definition: Digital transformation
The *use* of digital technology to enable a *significant* improvement in the realization of an organization's objectives that could not have been feasibly achieved by non-digital means

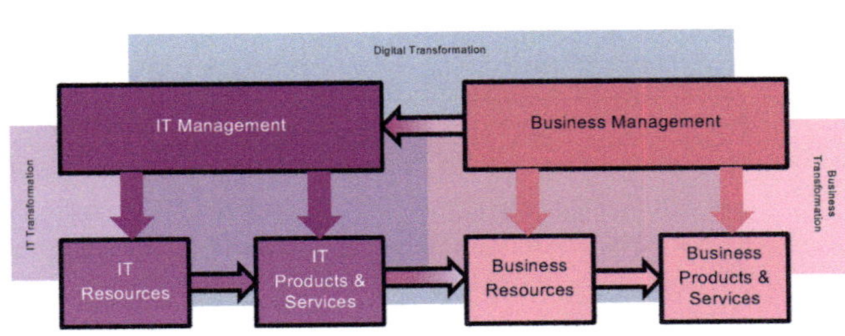

Digital products

A product is a configuration of an organization's resources designed to offer value for a consumer.

The product is provided to the consumer by means of services that enable the cocreation of value by both parties

Examples of digital products are online banking, ride sharing, e-commerce, e-tickets for events, photo and video storage and sharing, e-learning, airline flight booking and management, music streaming, smartphone apps, digital camera's etc

Definition: Digital product
A product is digital when digital technology plays a significant role in the resources that form the product, and the associated service interactions

Digital technology

Digital technology includes information technology (IT) and operational technology (OT)

Information technology is storing, retrieving, transmitting and manipulating data

- Only creates value when someone acts on a decision made using that data

Operational technology is using data to take physical action

- Car engine control, electricity grid, etc.

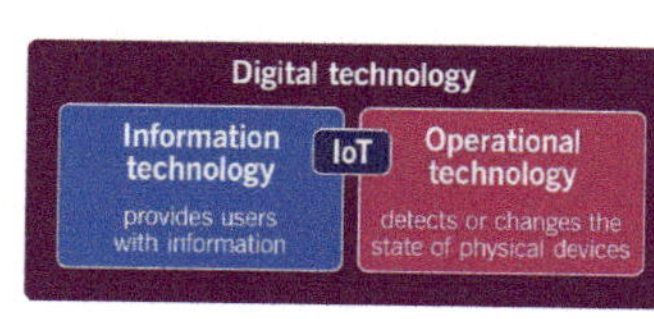

Definition: Digital technology
Technology that digitises something, or processes digital data. Digitisation is the process of transforming something from analogue form to digital form, in other words expressed in (binary) digits, or processing digital data. For example, text, sound and images. Digital technology refers to information technology (IT), and the parts of operational technology (OT) that have been digitised

Information technology

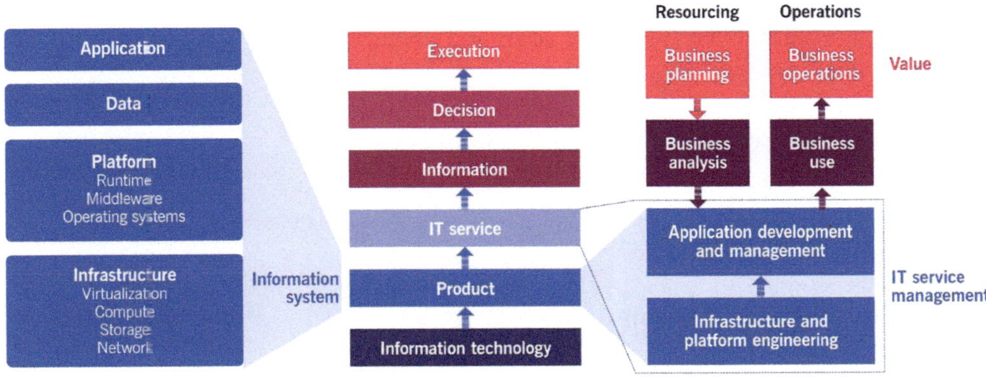

Definition: Information technology (IT)
The application of digital technology to store, retrieve, transmit, and manipulate data (data processing), often in the context of a business or other kind of organization

Operational technology

Operational technology differs from IT in that it uses digitised data as an internal means to a physical goal, rather than for the goal of making information available to users

Operational technology refers to physical devices such as valves and pumps in which digitised data is used to take physical action

The Internet of Things (IoT) gives the ability to connect OT devices to both other OT devices and to information systems

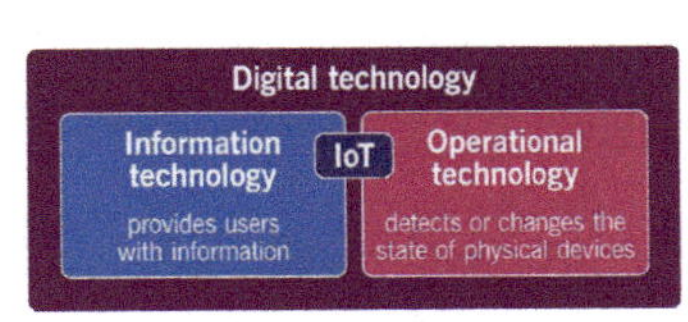

Definition: Operational technology (OT)
The application of digital technology for detecting or causing changes in physical devices through monitoring and/or control

When is transformation desirable and feasible?

Transforming the organisation to HVIT involves costs and risks

- But it could give a competitive advantage

It is a major change involving culture and other organisational aspects

- But it can be agile and incremental
- Some organisations need smaller improvements, not transformation to high velocity

Transformation is feasible when:

- There is a clear Return On Investment (ROI)
- Funding is available and secured
- Other required resources are available
- It fits the organisation's culture or the risk appetite

Five objectives to achieve

There are five objectives associated with digital products. The objectives are to achieve:

- **Valuable investments**
  – strategically innovative and effective application of IT
- **Fast development**
  – quick realisation and delivery of IT services and IT-related products
- **Resilient operations**
  – highly resilient IT services and IT-related IT-products
- **Co-created value**
  – effective interaction between service provider and consumer
- **Assured conformance**
  – to governance, risk and compliance (GRC) requirements

Figure 4.1: Balanced objectives

Note: There are potential conflicts between the objectives. For example, fast development could negatively affect resilient operations, so these objectives need to balanced

Calculating ROI for the objectives

The first objective, Valuable investments, represents the potential value of an investment. From an economic perspective it is possible to calculate a ROI as the difference between realised benefits and the costs incurred

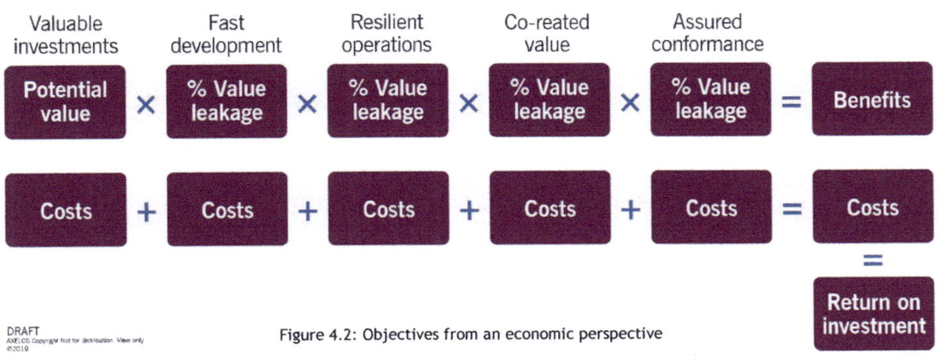

DRAFT
AXELOS Copyright Not for distribution View only ©2019

Figure 4.2: Objectives from an economic perspective

Objectives and the value chain

The five objectives closely relate to specific value chain activities:

- **Valuable investments**
  – mainly achieved by ***Plan***
- **Fast development**
  – mainly achieved by ***Design & transition***
- **Resilient operations**
  – mainly achieved by ***Deliver & support***
- **Co-created value**
  – mainly achieved by ***Deliver & support***
- **Assured conformance**
  – achieved by ***All activities***

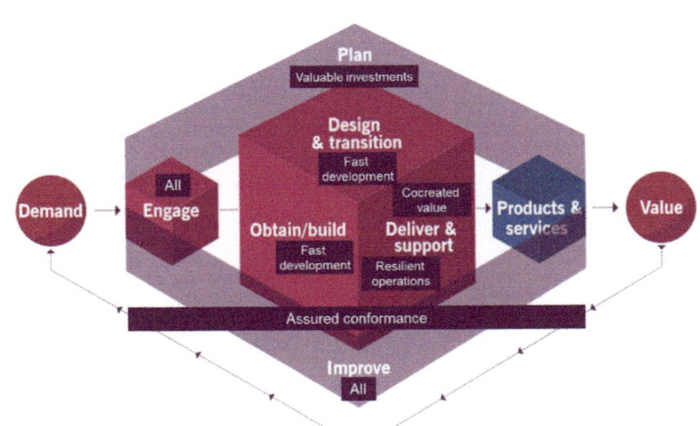

Figure 4.3: Objectives and service value chain activities

Engage and ***Improve*** activities also contribute to all of the five objectives

Techniques that characterise HVIT

ITIL 4 presents a selection of 24 techniques that characterise high-velocity IT environments

The techniques are grouped and described in relationship to one of the five high velocity IT objectives that are described below. Often, a technique supports multiple objectives

High velocity IT technique	Valuable investments	Fast development	Resilient operations	Co-created value	Assured conformance
Cost of delay	7, 11				
Minimum viable product	7				
Product ownership	7,9				
A/B testing	7				
Infrastructure as code		1			
Restrospectives		34			
Continual business analysis		16			
Continuous Integration/Continuous Deploy (CI/CD)		32, 33, 34			
Continuous testing		31			
Kanban		34			
Technical debt			7		
Chaos engineering			26		
Definition of done			31		
Version control			25, 34		
AIOps			28		
ChatOps			28		
Proactive monitoring			21		
Site Reliability Engineering (SRE)			15, 17, 18, 19, 21, 22, 33, 34		
Loosely-coupled architecture				1	
Service experience				9, 13, 16, 27, 28, 29	
DevOps Audit Defense Toolkit					3
Peer review					10
DevSecOps					3
Psychological safety					10

Table 4.1: Techniques that contribute to achieving the objectives, with reference to the supported objectives (shading) and practices (numbers)

General management practices	Service management practices	Technical management practices
1. Architecture management	15. Availability management	32. Deployment management
2. Continual improvement	16. Business analysis	33. Infrastructure and platform management
3. Information security management	17. Capacity and performance management	34. Software development and management
4. Knowledge management	**18. Change control**	
5. Measurement and reporting	**19. Incident management**	
6. Organizational change management	20. IT asset management	
7. Portfolio management	21. Monitoring and event management	
8. Project management	**22. Problem management**	
9. Relationship management	23. Release management	
10. Risk management	24. Service catalogue management	
11. Service financial management	25. Service configuration management	
12. Strategy management	26. Service continuity management	
13. Supplier management	27. Service design	
14. Workforce and talent management	**28. Service desk**	
	29. Service level management	
	30. Service request management	
	31. Service validation and testing	

Table 4.2: Practices (numbered)

Valuable investments

This objective is about market research and development of new products

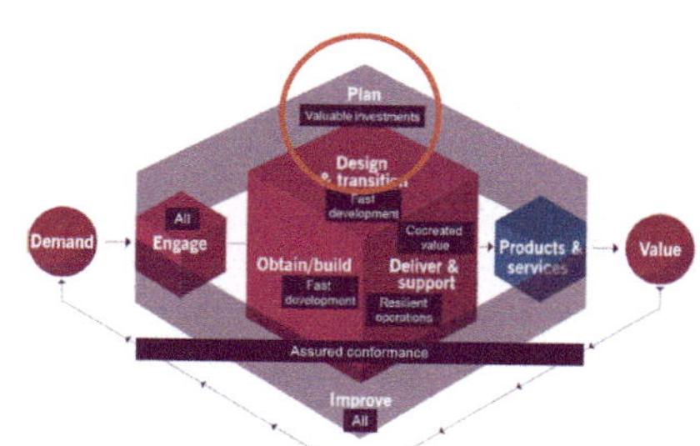

Scope of valuable investments: identify and justify digital investments that contribute significantly to business strategy

The potential value of digital investments is determined at this stage

- Utility determines potential value
- Warranty ensures that potential value is not adversely affected

Continually evaluate and re-evaluate products and services to identify better alternatives

Fast development

This objective is about enabling faster ROI by fast development

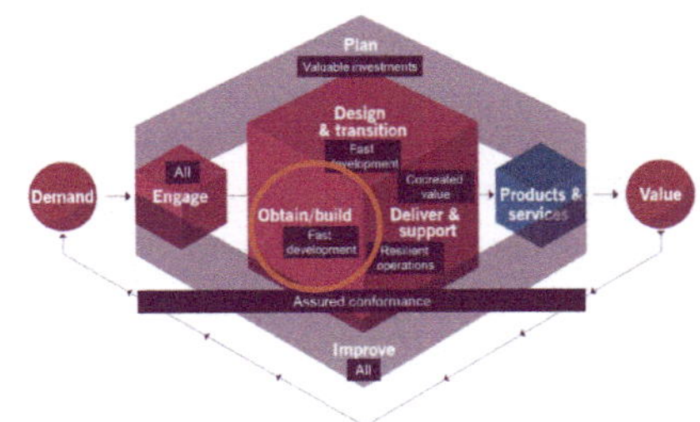

Scope of fast development: realize new and improved digital products and services frequently, quickly and reliably

In general, the sooner digital products are delivered, the sooner the value can be realized

Breaking down the product into a series of deliveries of small increments can increase speed

In addition to delivering quickly and frequently, delivery must be reliable and able to restore

Fast development has no intrinsic value. Value comes from the use of the product or service

Fast development

Because of diametrically opposed targets – make changes very quickly versus make changes very carefully – development and IT operations are in a state of constant battle

By reducing the size of change, the risk of disruption is reduced. Smaller changes means that change happens more frequently. By changing more frequently, the organization's capability to change is improved. Increased change capability leads, in turn, to lower risk of disruption

The DevOps community has embraced this way of thinking and developed appropriate practices and tools

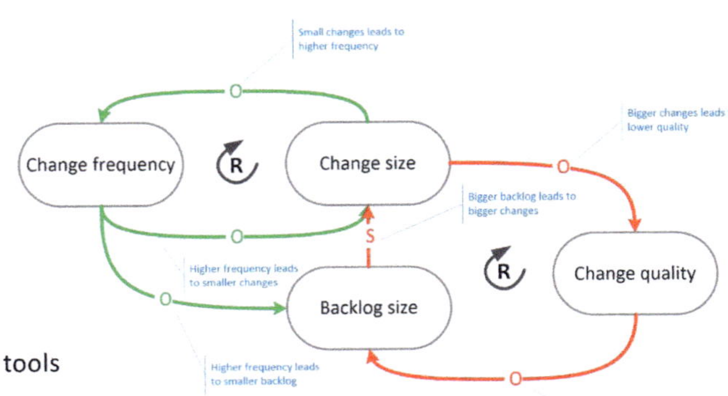

Figure 4.9: Effect of the size of changes

Resilient operations

This objective and scope is about ensuring digital products and services are available for use whenever needed

Resilience applies to all parts of the systems stack and also to the organization that manages these component parts

Failsafe systems are an illusion. The emphasis is on restoring service quickl

Resilient operations can be measured in terms of availability, performance, and security

It adds nothing to the potential value of an investment: it 'just' ensures that the potential can be realized

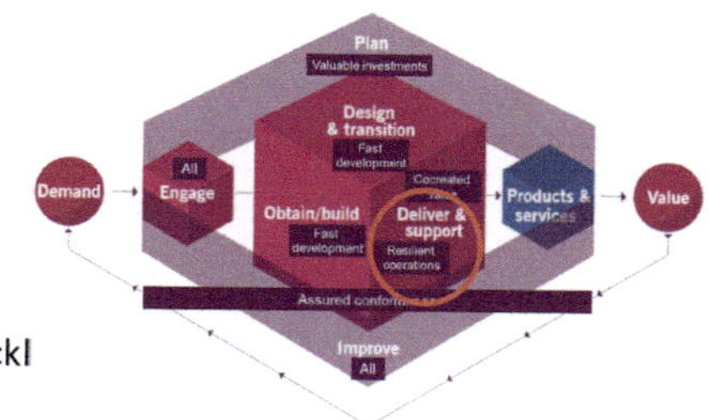

Co-created value

Co-created value is about the service consumer using the service provider's products and services effectively, and benefitting from their utility and warranty

Effective interaction between service provider and consumer is essential and good understanding of digital products in terms of business process is key

Scope of cocreated value: cocreate value from digital products in close collaboration between service provider and consumer

Co-created value refers not only to the value for the service consumer, but also for the provider and any other stakeholders

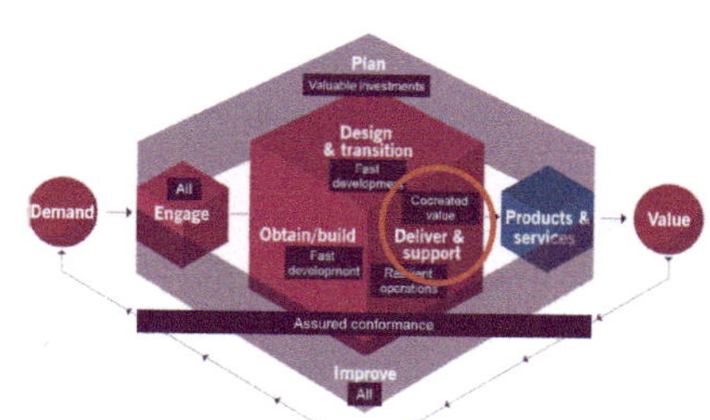

Assured conformance

This objective is about conformance and assurance

Scope of assured conformance: assure that service provision and service consumption comply with corporate and regulatory directives with respect to governance, risk and compliance (GRC)

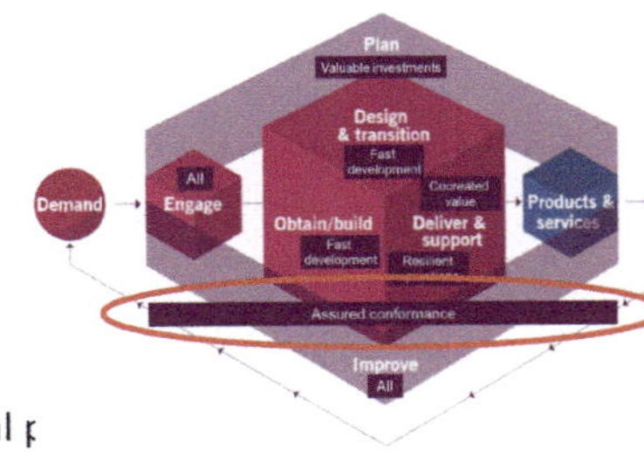

High velocity is often associated with taking risks, and from a commercial p
to take risks. Nevertheless, risks must be justified and there are internal rules and external regulations that must be complied to

Assured conformance contributes to the peace of mind of the people who are accountable for, and affected by governance, risk and compliance issues

How HVIT relates to...

- The four dimensions of service management
- The ITIL service value system
- The service value chain
- The digital product lifecycle

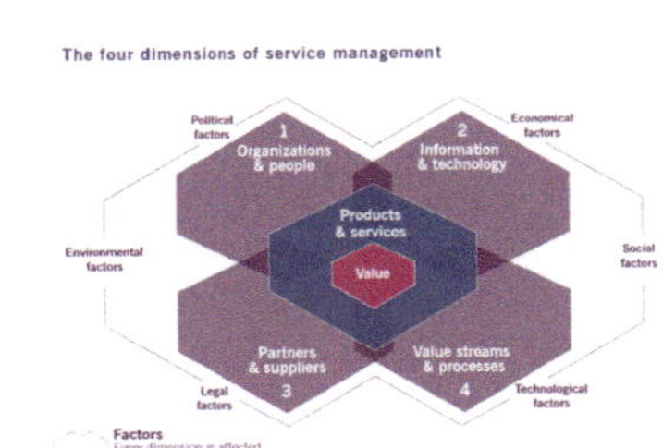

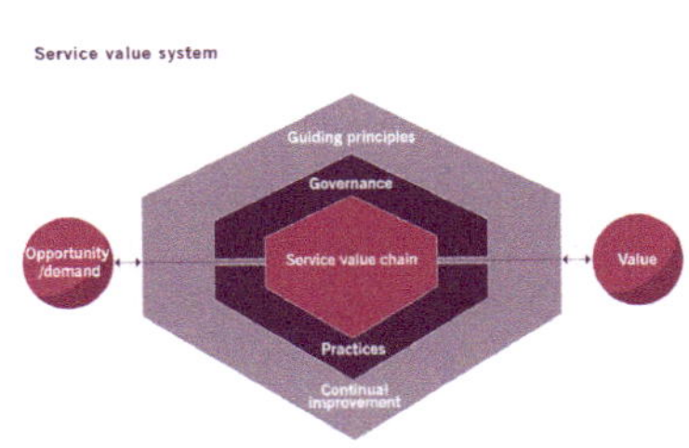

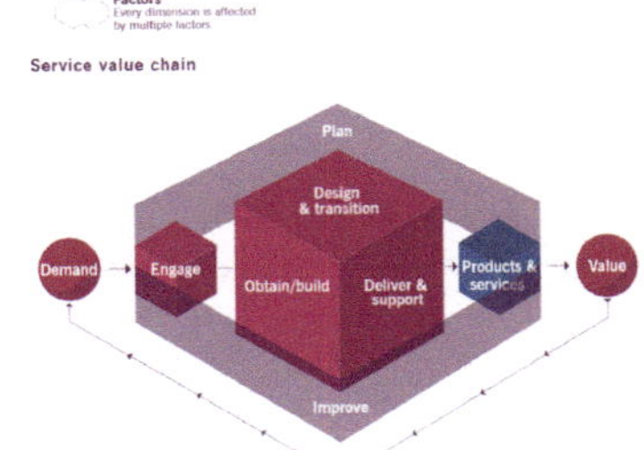

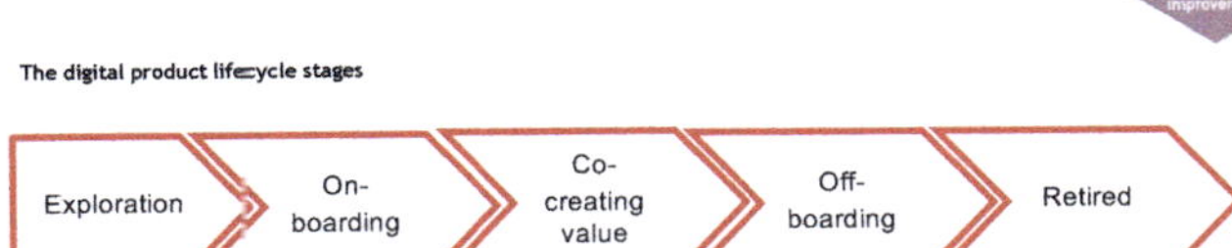

The four dimensions of service management

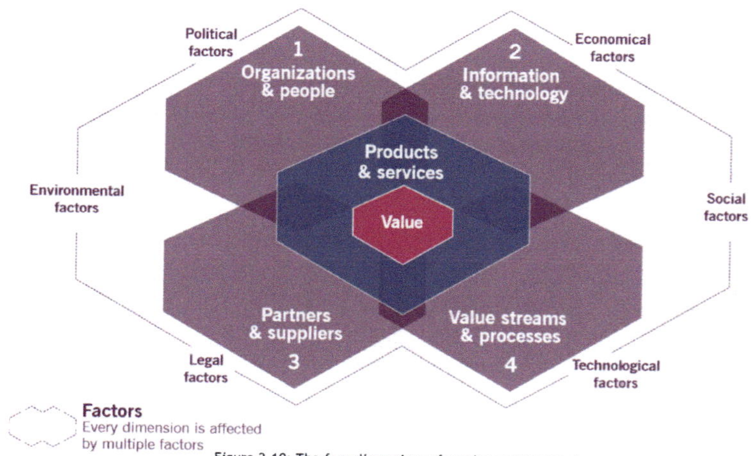

Figure 2.10: The four dimensions of service management

HVIT and organisations & people

If IT is an integral part of the organization's products and services, it is likely that the IT function is an integral part of the lines of business

Although there may be a centralized IT service centre for non-differentiating digital technology

There will often be multifunctional product/service-based teams, with both business-oriented and IT-oriented team members and less focus on formal agreements, such as SLA's

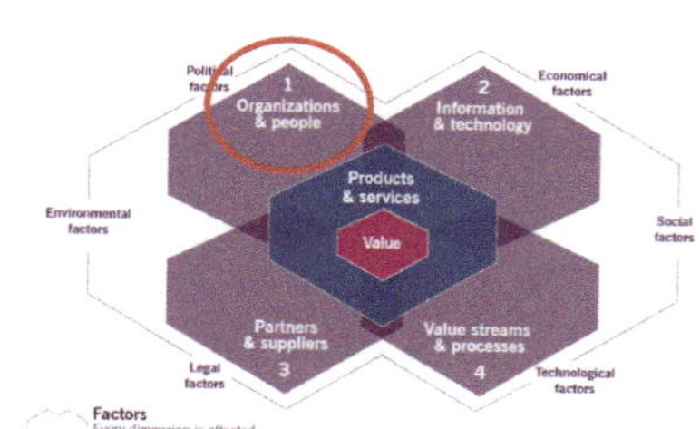

Figure 2.10: The four dimensions of service management

This is not only good for communication, but also for building a better and common understanding of the business context in which digital technology is used

HVIT and information & technology

Higher demands are placed on both the information and technology that digitally enable organizations, and the information and technology (IT tooling) that support the IT processes

Practitioners should be increasingly aware of the data that they have at their disposal, and make better use of it

It is common to find that information is created and used in real time and at times spontaneously. Freely flowing information is a characteristic of a high-trust organization

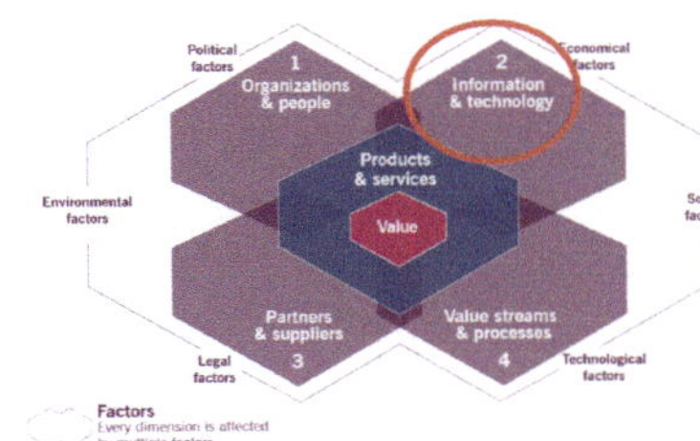

Figure 2.10: The four dimensions of service management

The expected extreme growth of AI and machine learning will place higher demands on well-managed information and knowledge

HVIT and partners & suppliers

High velocity IT environments often make extensive use of cloud-based infrastructures, platforms and other services

Public cloud-based services are usually high quality and affordable, but the individual consumer has little or no influence on the provider

It is therefore crucial to analyse the dependencies and take appropriate action in terms of contracts and SLA's

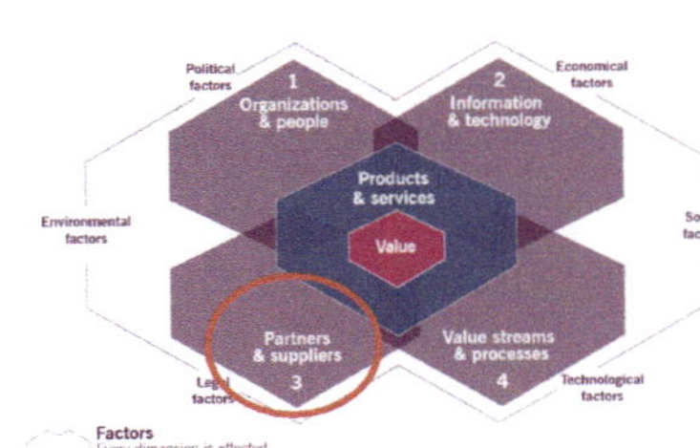

Figure 2.10: The four dimensions of service management

Important to consider whether the external service provider has a similar way of working to their customer as it is difficult to integrate parties with different ways of working into the same value stream

HVIT and value streams & processes

It may be better to create a unique value stream for each digital product or service

This may be less efficient than a standardized and centralized single value stream that serves multiple products and services, but the benefits in terms of effectiveness many times will outweigh the costs

Some kind of standardisation may be required for scalability, so the trade-off needs to be carefully considered

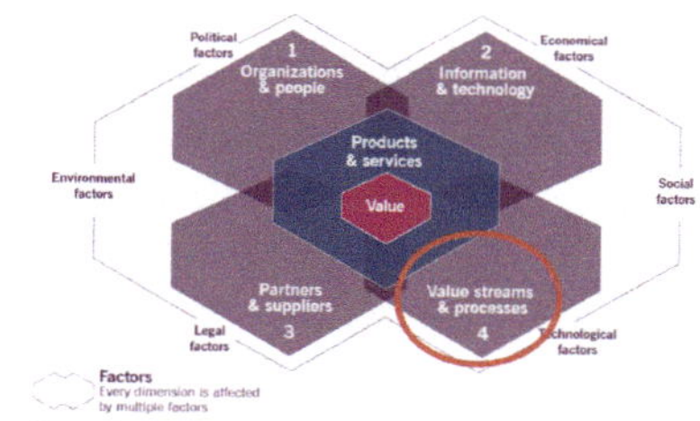

Figure 2.10: The four dimensions of service management

Because processes are predetermined, they are applicable to situations that are predictable. In unpredictable circumstances, a case-based approach is often more effective

External factors

Political, economic, social, technological, legal and environmental factors (PESTLE) influence an organization's choices for markets, products and services, and resources and activities

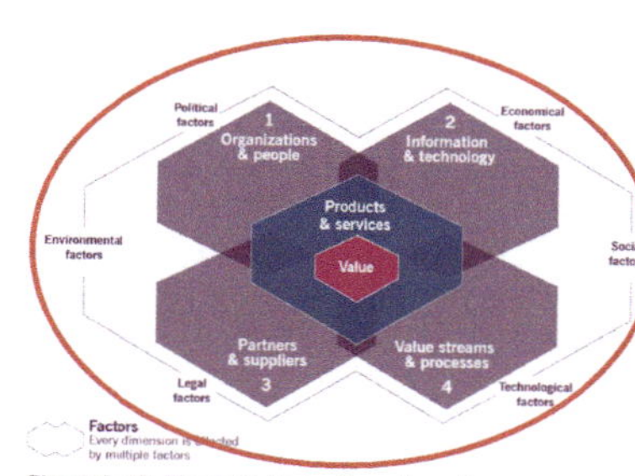

Figure 2.10: The four dimensions of service management

HVIT environments are characterized by relatively high degrees of volatility, uncertainty, complexity and ambiguity (VUCA):

- **Volatility** - The nature and dynamics of change, and the nature and speed of change forces and change catalysts
- **Uncertainty** - The lack of predictability, the prospects for surprise, and the sense of awareness and understanding of issues and events
- **Complexity** - The multiplex of forces, the confounding of issues, no cause-and- effect chain and confusion that surrounds organization
- **Ambiguity** - The haziness of reality, the potential for misreads, and the mixed meanings of conditions; cause-and-effect confusion

The ITIL Service Value System

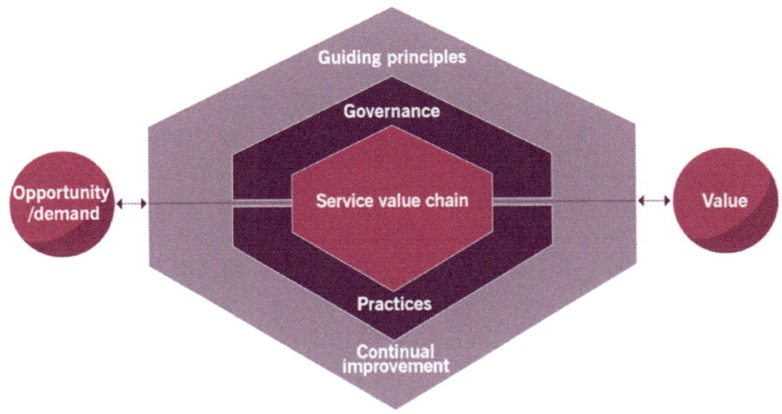

Figure 2.11: The ITIL Service Value System

The ITIL service value system

How do HVIT organisations adapt the ITIL service value system to their specific needs?

Are all 5 components of the SVS as important in HVIT as in a more "traditional" IT environment?

Give a few examples of what differences you have observed in your experience of HVIT relating to one or more of the components of the SVS

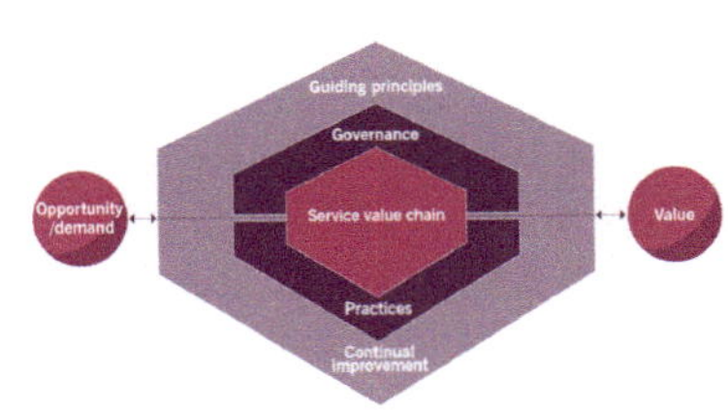

Figure 2.11: The ITIL Service Value System

Note: Understanding the SVS is foundational for the overall understanding of ITIL 4. It is also critical to apply the understanding of these concepts to your own real-life experience!

Governance and Management

Governance and management are as important in HVIT as in any other context

Separate governance from management!

Governance are the means by which an organisation is directed and controlled. This is done by evaluating, directing and monitoring (EDM)

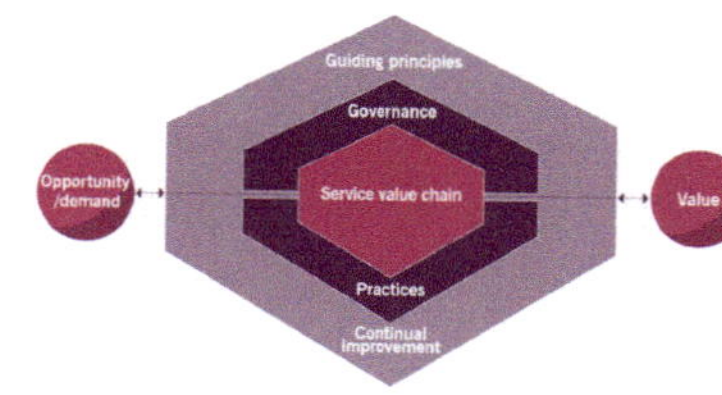

Figure 2.11: The ITIL Service Value System

Management is responsible for planning, building, organising and improving the organisation

In digitally enabled organizations with lines of digital business, the responsibility for digital technology is an integral part of the line of digital business. IT is not a separate unit

Managing for agility and resilience

Practitioners operate within a governance and management framework. They understand the applicable constraints and know how to act within that framework

High velocity IT environments are often highly unpredictable, meaning that predetermined instructions are neither feasible nor desirable

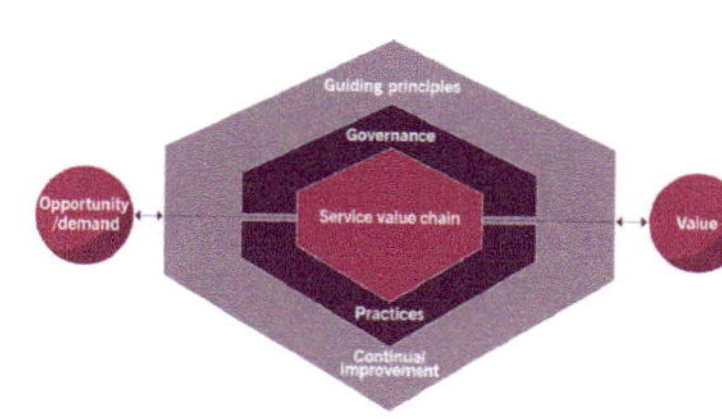

Figure 2.11: The ITIL Service Value System

The high velocity IT practitioner therefore has to exercise judgement on the job. In order to do this effectively, the practitioner has to understand the thinking behind the constraints

This is a major role of the manager in the high velocity IT environments: to provide context and to enable the practitioner to take charge

The service value chain

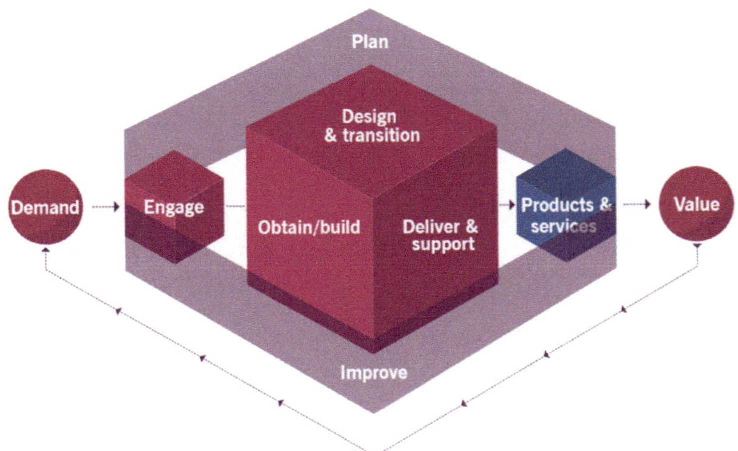

Figure 2.12: The service value chain

The service value chain

How do HVIT organisations adapt the service value chain to their specific needs?

Are all 6 key activities of the service value chain as important in HVIT as in a more "traditional" IT environment?

Give a few examples of what differences you have observed in your experience of HVIT relating to the service value chain and one or more of its key activities

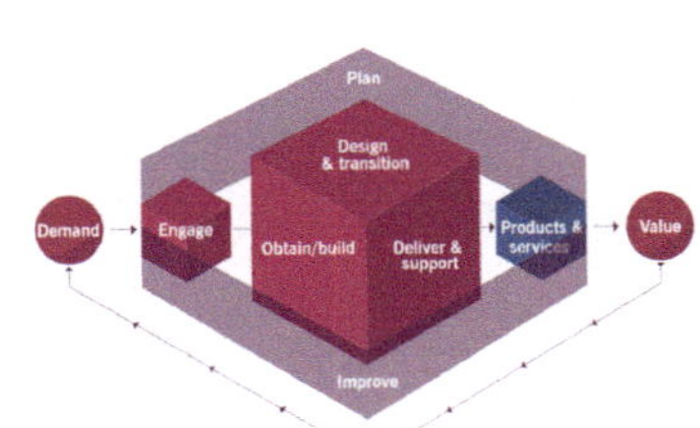

Figure 2.12: The service value chain

Note: As the 'operating model' of ITIL 4 the service value chain acts as the main environment for any value stream developed to provide value to the service consumers

What is an operating model?

An operating model is the 'back end' of the organisation's business model

Just like the business model, the operating model is a representation of the building blocks of an organization and the relationships between them

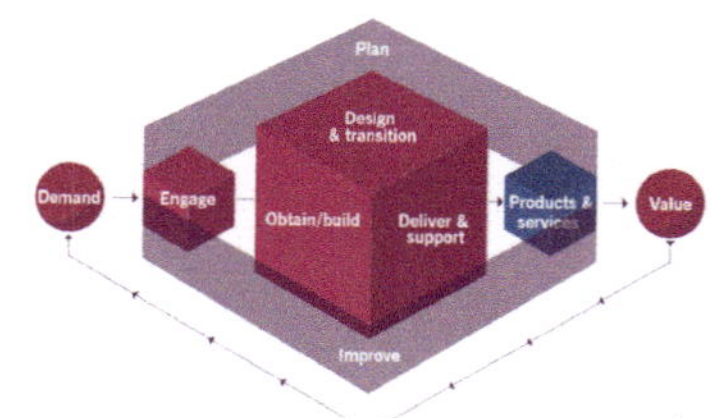

Figure 2.12: The service value chain

It can be used as a description of the current state of the organization, or as a future state design (a target operating model)

Definition: An operating model is a conceptual or visual representation of how an organization cocreates value with its customers and other stakeholders, as well as how the organization runs itself

Digital or HVIT IT operating models

'Digital operating model' refers an organisation or part of an organisation where digital technology plays a major role in the co-creation of value

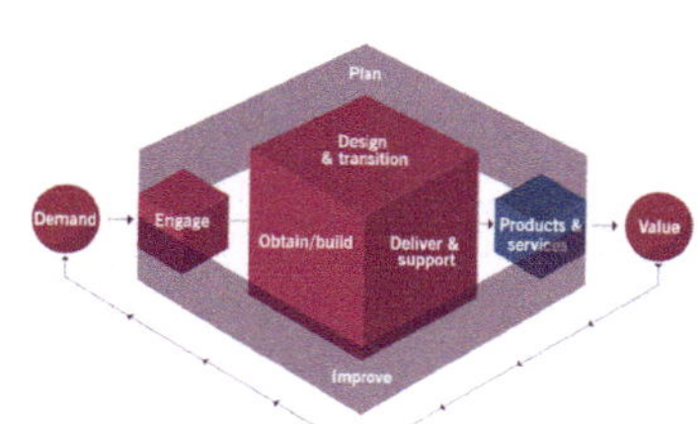

Figure 2.12: The service value chain

Operating models for digitally enabled organizations are characterized by their focus on:

- Dedicated value streams for each of the various products and services
- Cocreational culture that fosters high performance and continual improvement
- Permanent product/service-based teams over temporary project teams
- Automation of IT processes, including infrastructure as code

Definition: A digital operating model or 'HVIT operating model' is an operating model where digital technology plays a major role in the co-creation of value

ITIL building blocks

From an operating model perspective, ITIL comprises interrelated organisational building blocks that revolve around the service value streams

A service interaction between consumer and provider is a combination of three kinds of interactions:

- Service actions
- Access to resources
- Transfer of goods

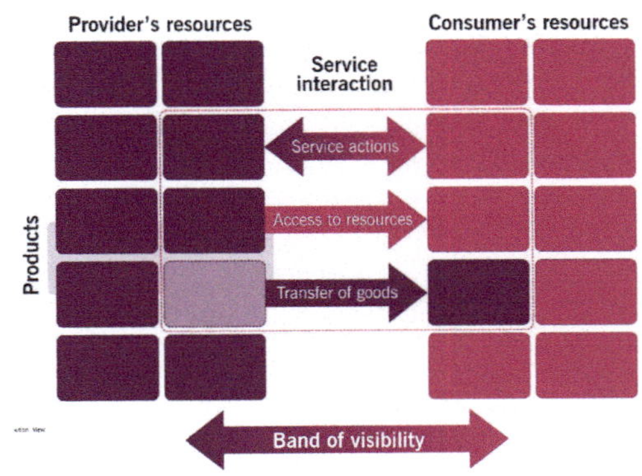

Figure 2.9: Service interaction

This is called the band of visibility. The width of the band of visibility is crucial to the service experience and the effectiveness of the service interaction. A delicate balance has to be struck here

Consumer engagement with multiple service providers

The service consumer often engages with multiple service providers to get their jobs done

The service value chain is for the service provider

There are interactions with the customer and consumer but the activities are intended to guide the service provider

The service consumer has a complementary value chain, that interacts with the service provider's value chain

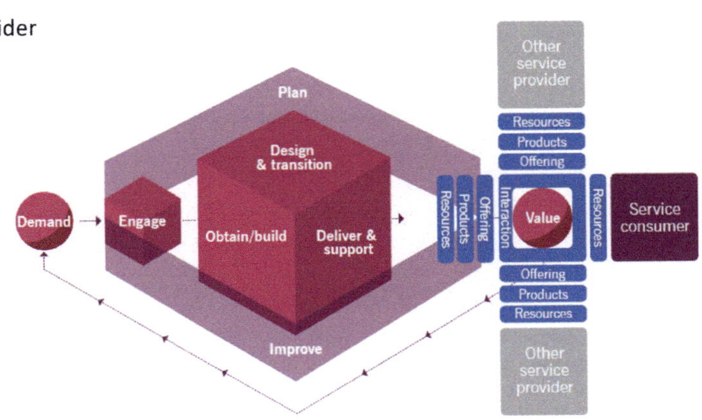

The use of service value chains

'Digital operating model' refers an organisation or part of an organisation where digital technology plays a major role in the co-creation of value

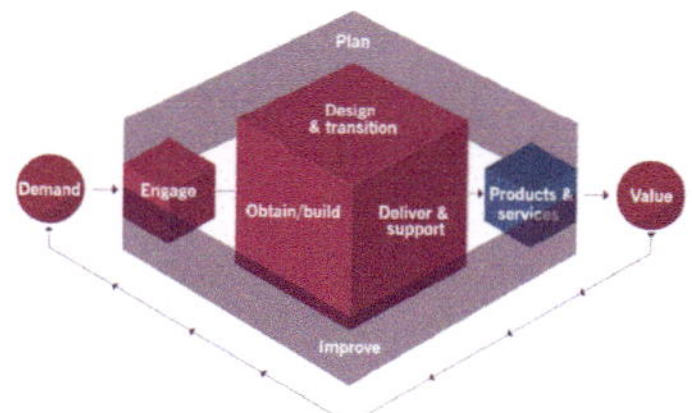

Figure 2.12: The service value chain

Operating models for digitally enabled organizations are characterized by their focus on:

- Dedicated value streams for each of the various products and services
- Cocreational culture that fosters high performance and continual improvement
- Permanent product/service-based teams over temporary project teams
- Automation of IT processes, including infrastructure as code

Definition: A digital operating model or 'HVIT operating model' is an operating model where digital technology plays a major role in the co-creation of value

DevOps & the service value chain

In high velocity IT environments, the concept of 'continuous' is often used. This refers to fast iterative cycles of activities, enabled by short feedback loops

The DevOps community often illustrates this with a figure of a continuous loop with eight connected application development and IT operations activities, as shown in the marked part of the figure

Dev and Ops activities are enabled by infrastructure and platforms for coding, testing, deployment, production etc.

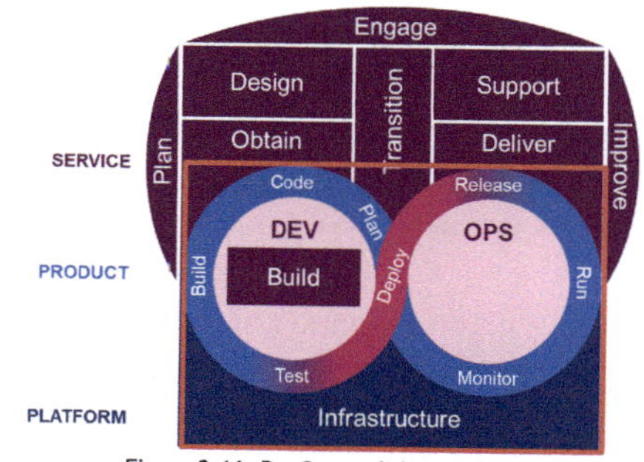

Figure 2.16: DevOps and the service value chain

The service value chain activities can be used to unite this DevOps perspective with an ITIL perspective in IT service management terminology

The service consumer & the service value chain

Another example of a high velocity IT application of the service value chain, is to illustrate how closely service provider and service consumer interact

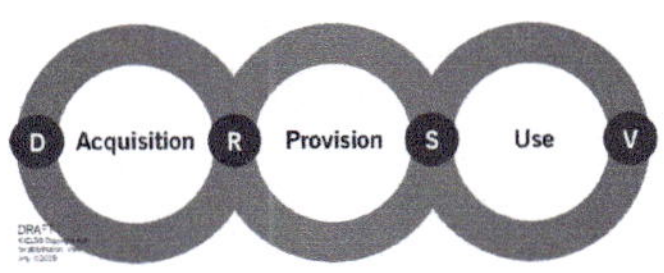

Figure 2.17: The service consumer's perspective (D=demand, R=requirements, S=service, V=value

As with the DevOps example, the combined perspectives can be used for explanation and discussion purposes

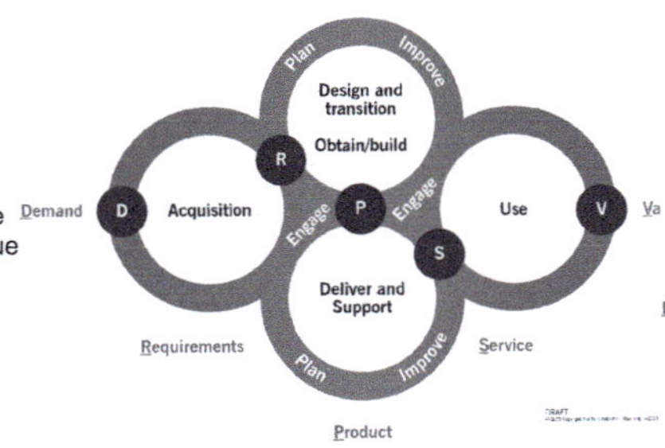

Figure 2.18: Interacting service value chains

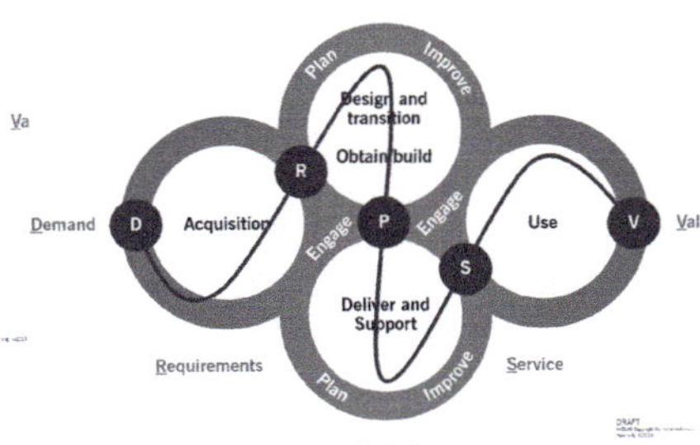

Figure 2.19: Example service value stream referring to service value chain activities

How these concepts all connect...

Service value streams tie these concepts together and gives context...

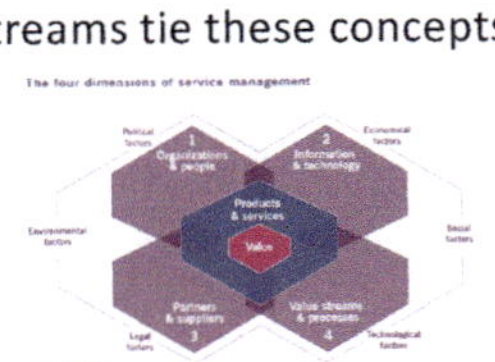

Figure 2.10: The four dimensions of service management

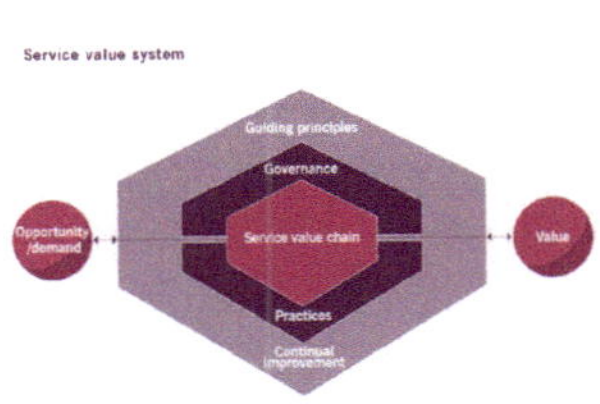

Figure 2.11: The ITIL Service Value System

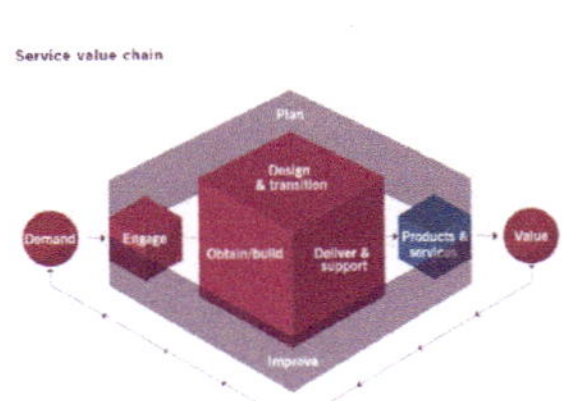

Figure 2.12: The service value chain

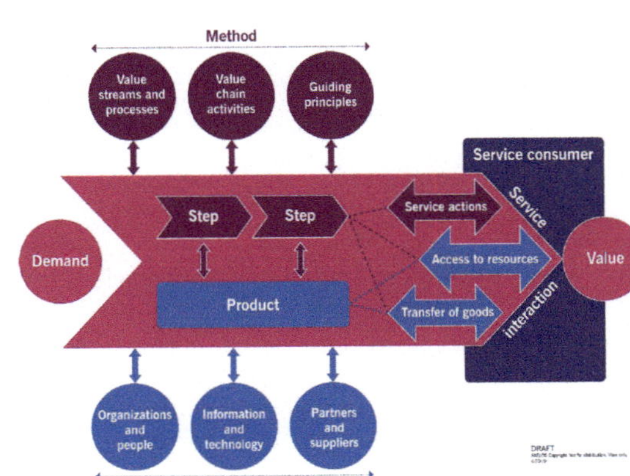

Figure 2.13: The service value stream in context

Governance, execution and improvement of value streams

Service value streams need to be governed, managed and improved

In figure 2.14, governance, execution and improvement are regarded as three interacting dimensions

Governance governs both execution and improvement, and improvement improves both execution & governance

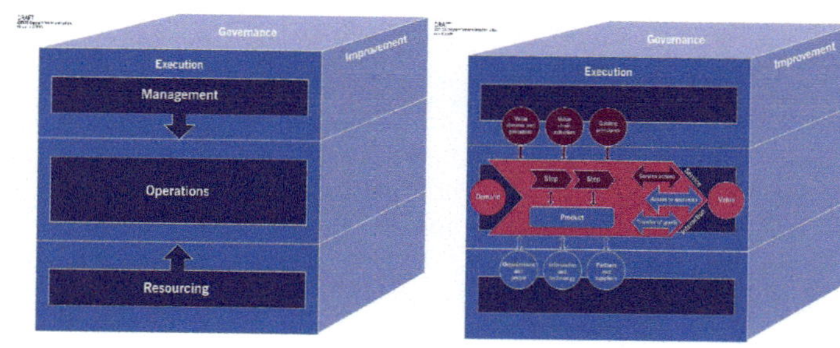

Figure 2.14: Service value stream positioned with respect to governance, execution and improvement

The digital product lifecycle

Service consumer and service provider have different perspectives on digital products

They each have their own lifecycles, that overlap during the period of engagement

For the service provider, the lifecycle is as long as there are potential customers for the product

For the service consumer, the lifecycle is as long as the product is used

Strictly speaking, it is a product *use* lifecycle

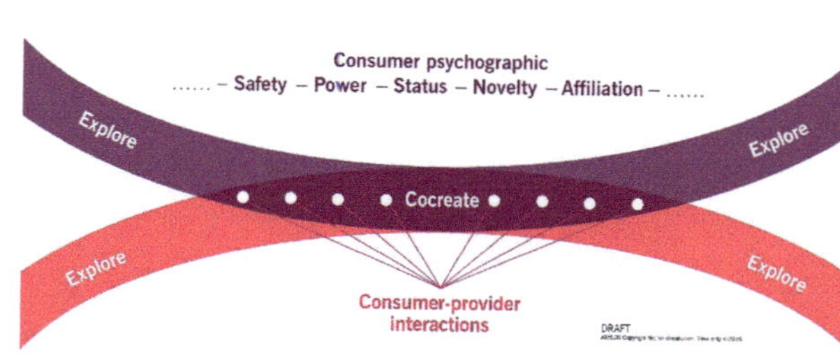

Figure 2.20: Digital product lifecycle from two perspectives - consumer and provider

The customer journey model

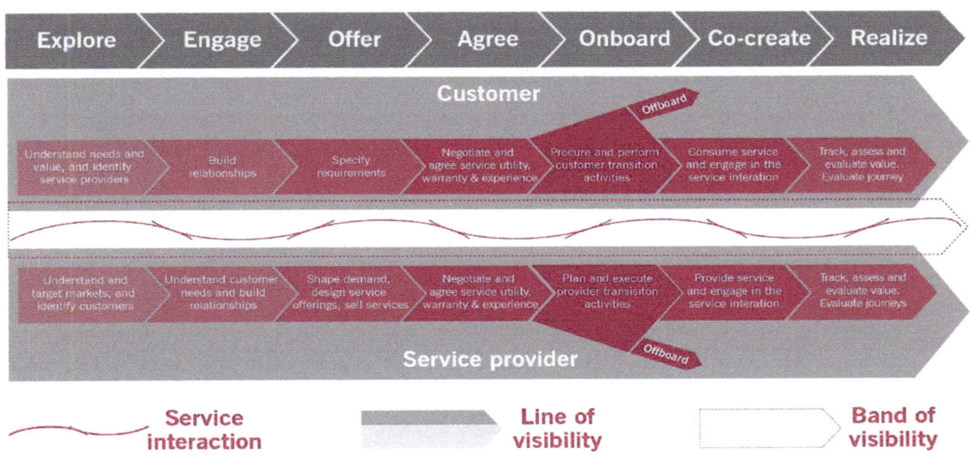

Figure 2.21: Customer journey model

The digital product lifecycle stages

The digital product lifecycle can be simplified to the following stages:

- Exploration
- Onboarding
- Co-creating value
- Offboarding
- Retired

From a consumer's perspective the digital product lifecycle can be seen as a series of engagements with different providers and their services

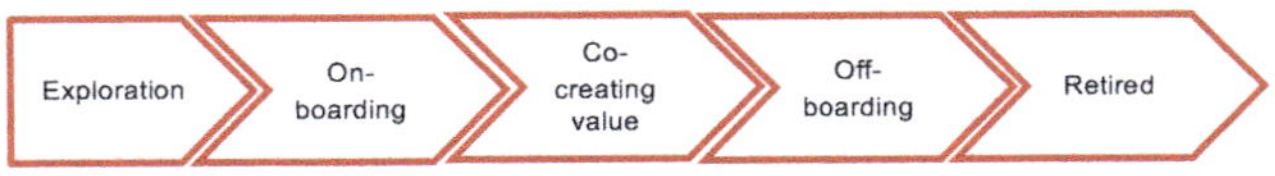

Figure 2.15: Digital product lifecycle stages

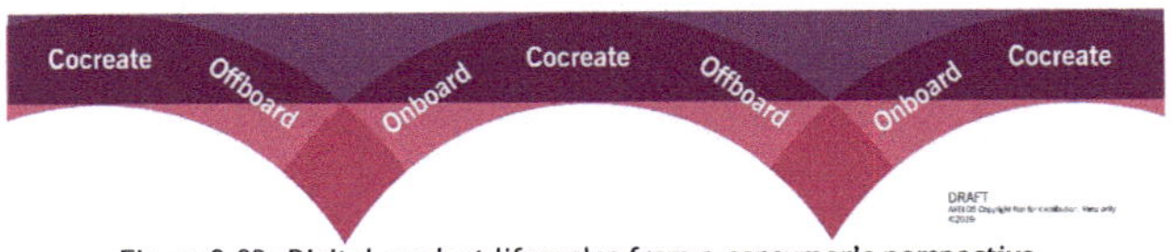

Figure 2.22: Digital product lifecycles from a consumer's perspective

Culture supporting & enabling HVIT

This chapter explores the kind of culture that supports and enables high velocity IT work

It proposes five key behaviour indicators that reflect organizational needs as well as practitioner's aspirations to work in a rewarding environment. These are supported by a combination of guiding principles, models and concepts that inform "how we do things around here" – in other words, culture

We will later on go through some examples of these principles, models and concepts

These are grouped into three categories:

- **Purpose**
- **People**
- **Progress**

Deal with ambiguity & uncertainty

This reflects the volatile and ambiguous nature of the typical digitally enabled organization

In these environments, where failsafe systems are often an illusion, we favour experimentation and are not afraid to fail safely and mindfully

We are not scared by not knowing. Things aren't perfect and they never were, and we have techniques to deal with that

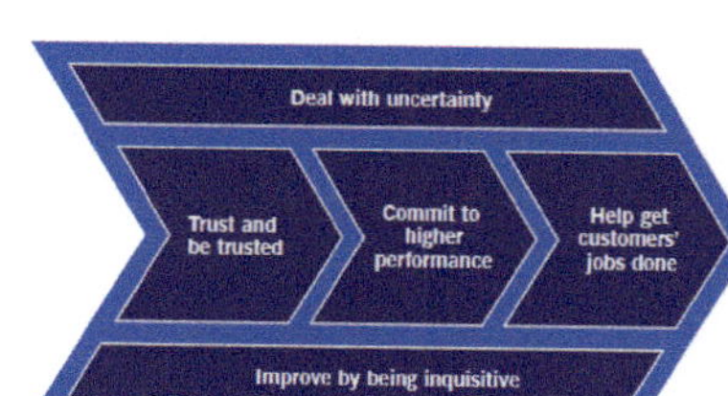

Figure 3.1: Key behaviour indicators

Trust and be trusted

This is not only about respecting people's professional skills and trusting them to take the right decisions, but also about being considerate to fellow human beings, acknowledging diversity and being inclusive

People also deserve honest – but considerate – feedback

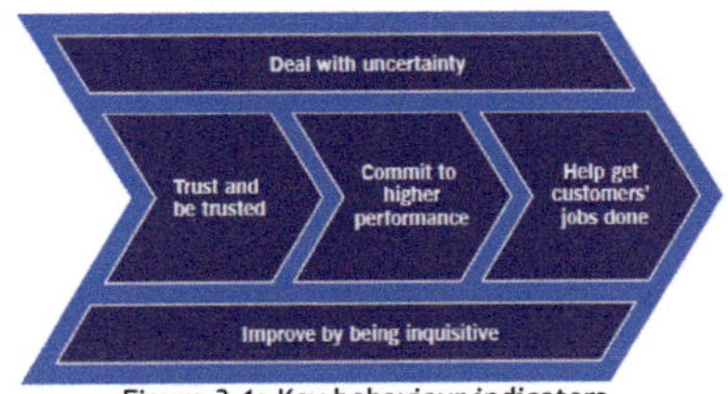

Figure 3.1: Key behaviour indicators

Keep raising the bar

This refers to never being satisfied with the status quo, because even if it is satisfactory today, it won't be tomorrow

Customers and users are always pleasantly surprised when you've taking the initiative to make improvements. And there's always an improvement to be made, however small

It's an attitude. You are encouraged to show leadership by spotting improvement opportunities and contributing to following them up

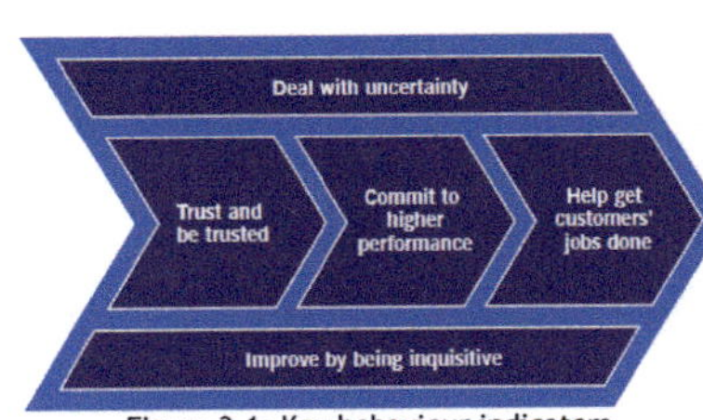

Figure 3.1: Key behaviour indicators

Help get customers' jobs done

This is the essence of every organization. There are, of course, other stakeholders, but customers (or citizens) are the organization's raison d'être

This is the generous act of helping somebody solve their problems and become who they seek to become. It recognizes the psychographics of the customer – how the customer wants to feel before, during and after using the digital and physical products and services

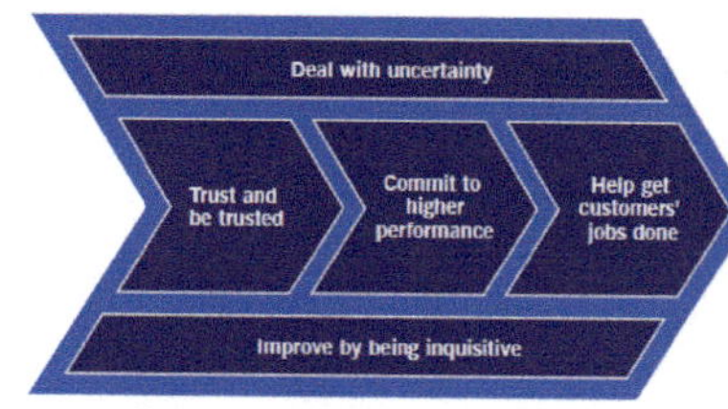

Figure 3.1: Key behaviour indicators

It involves making a bold and honest assertion: "for people who believe A and want B, our offer will bring C". If they don't believe A or want B, then "sorry, it's not for you"

Commit to continual learning

This underpins the other areas. The root cause of everything is ignorance

Either that somebody didn't have the right information when they had to act, or when they were creating the 'system' in the first place. Ignorance can also refer to believing that situations are knowable, which leads us back into accepting ambiguity and uncertainty

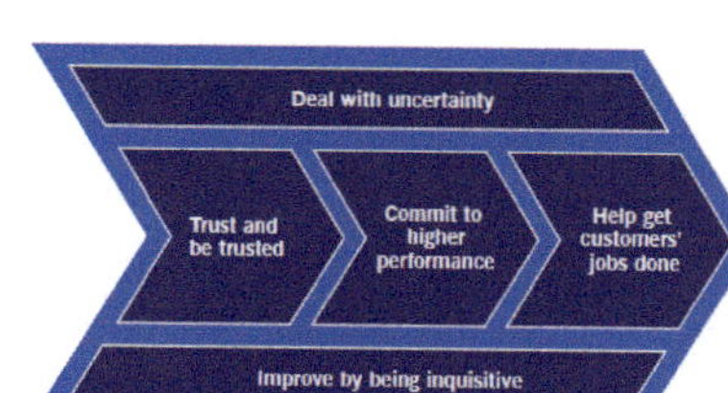

Figure 3.1: Key behaviour indicators

Mapping of behaviours & principles

	Accept ambiguity and uncertainty	Trust and be trusted	Continually raise the bar	Help get customers' jobs done	Commit to continual learning
PURPOSE					
Ethics	X	X	X	X	X
Design thinking				X	
PEOPLE					
Reconstructing for service agility			X		
Integration of duties		X	X		
Safety culture		X			
Stress prevention		X			
PROGRESS					
Complexity-based heuristics	X				
Lean culture		X	X		X
Toyota Kata	X				X
Continual improvement model			X		X
ITIL guiding principles	X	X	X	X	X

Fundamental concepts for HVIT

We will now look at several principles, models and concepts for delivering HVIT

This is not a complete list, but cover a broad set:

- Ethics
- Design thinking
- Safety culture
- Lean culture
- Toyota Kata
- Lean / Agile / Resilient / Continuous
- Service-dominant logic
- Complexity-based heuristics

Ethics

As mentioned in the introduction, the societal, political and economic impact of IT is unprecedented

For better and for worse. Digital organizations – organizations in which IT drives the business rather than just supports it – therefore have an increasingly strong moral obligation to consider how they apply IT, beyond their direct economic interests

> **Definition: Ethics**
> A system of principles which defines what is good for individuals and society

Safety culture

Digitally enabled organizations in particular are under pressure to continually improve their performance under changeable market conditions. Technological progress means that their products have increasingly significant economic, societal and political impact

And failure, with disastrous consequences. The organizational challenge is to foster an effective set of shared beliefs, perceptions and values in relation to risks

In a safe culture, people feel trusted and valued. They are therefore more likely to point out risks than when they fear that this would damage their reputation and position

> **Definition: Safety culture**
> A climate in which people are comfortable being (and expressing) themselves

Lean culture

Lean is a balance of striving for standardisation and predictability to avoid errors, while fostering a cultural of calculated risk-taking, curiosity and inquiry, based on a foundation of trust and respect

To be effective, resilient and adaptable, a high-velocity IT environment must be built on a bedrock of Lean culture

Definition: Lean culture
A way of working where trust, respect, curiosity, inquiry, playfulness and intensity all co- exist in a plasma of learning and discovery

Toyota Kata

Disciplined scientific thinking with incremental improvements reduces wrong conclusions

Practice and coaching help us form and sustain new habits, after which we can start improvising

The four-step improvement kata is based on five questions:

1. What are we trying to achieve?
2. Where are we now?
3. What obstacle is now in our way?
4. What's our next step, and what do we expect?
5. When can we see what we've learned from taking that step?

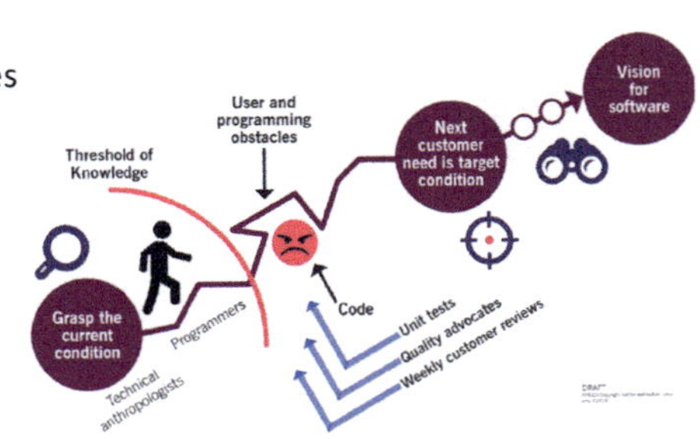

Figure 3.3: The Improvement Kata
[source Rother, Toyota Kata book]

Definition: Toyota Kata
A mental model and behaviour pattern for scientific thinking and routines for practice and coaching

Lean/Agile/Resilient & Continuous approaches

Lean, agile, resilient, continuous and co-creational are dominant characteristics of common high velocity IT approaches

These approaches are technical in nature, focusing more on the tangible parts of information systems (the product), but many of the principles can also be applied to the more abstract concept of IT service

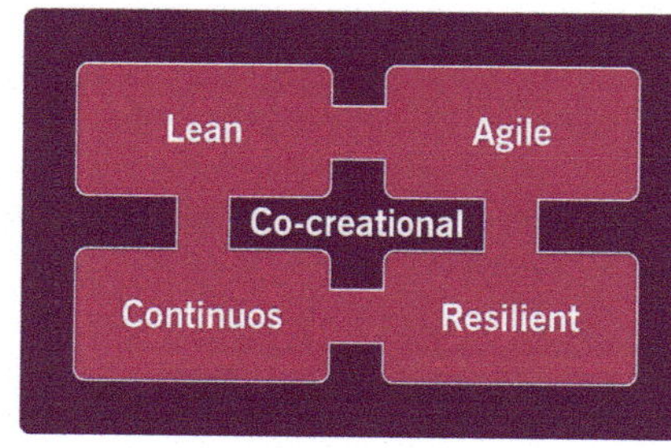

Figure 2.6: Lean/Agile/Resilient/Continuous/Co-creational approaches

Figure 2.8: a value stream with three workstations

Co-creational approach: Service-dominant logic

Service science defines service as the application of resources (including competencies, skills, and knowledge) to make changes that have value for another

A core concept in service science is service-dominant logic

Service-dominant logic regards service as the process of doing something for and with another party. Value creation is a collaborative process

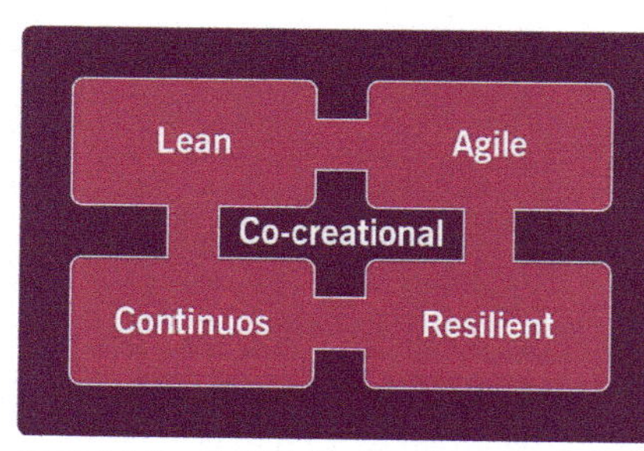

Figure 2.6: Lean/Agile/Resilient/Continuous/Co-creational approaches

Definition: Service-dominant logic
A mental model of (economic) exchange in which organizations7 cocreate value by applying their competences and other resources for the benefit of another

Design thinking

In digitally enabled organizations, the quality of the digital products and customer experiences is paramount

An understanding of design thinking gives practitioners the ability to contribute – together with co-workers and customers – to creating better digital products and customer experiences, and helping to complete jobs for customers effectively

Design thinking is particularly useful when addressing 'wicked problems', in other words when there are unarticulated needs and conflicting hypotheses

Although it is continually evolving, it's not new, but it has recently attracted more interest

Definition: Design thinking
The set of cognitive and practical processes by which design concepts are developed

Complexity-based heuristics

In order to work effectively, you have to understand the nature of the environment – or 'system' – in which you work

Systems thinking in general, and complexity thinking in particular, help make sense of the system guidance to effective approaches

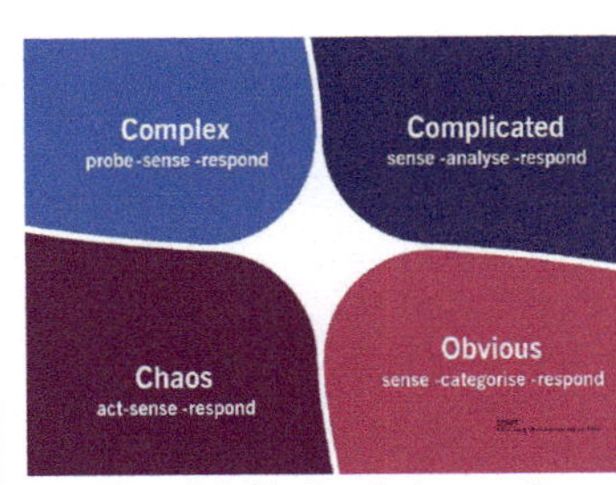

Figure 3.2: The Cynefin framework

Definition: Complexity thinking
The understanding that some systems are unpredictable and can only be influenced by experimentation with attractors, repellers, and modifications to the boundaries that constrain the system's parts

Definition: Complex adaptive systems
Systems that adapt in, and co-evolve with, a changing environment, resulting in (1) behaviour that is not predicted by the behaviour of parts of the system, and (2) the inability to examine the system in isolation from the other systems in its environment

Summary

We have just talked about:

- ✓ Make the right digital investments
- ✓ Realize and deliver digital products and services quickly
- ✓ Provide digital products and services that are highly resilient to disruption
- ✓ Ensure that the service consumer realizes value from the digital products and services
- ✓ Assure conformance of activities with governance, risk and compliance requirements

Q: What does a digital organisation have to do to create or maintain a competitive advantage?

A. Market and sell existing products and services to new customers

B. Deliver innovative new products and services quickly and reliably

C. Provide multiple, integrated support channels for products and services

D. Change the way that the IT function develops products and services

Q: An organisation has a 'failure is not an option' culture. As a result, it is common for employees to hide mistakes? Which would BEST help this organisation adopt a more open and trusting environment where people don't blame each other for failures?

A. Safety culture

B. Relationship management

C. Service-dominant logic

D. Cynefin

investment strategies. The IT staff work long hours under pressure to manage the algorithms required. An increasing number of errors are occurring, and staff are disagreeing and blaming each other. What approach should be taken

A. Reinforce trust behaviour by fostering a safe environment in which to voice opinions and resolve issues

B. Introduce chaos engineering to proactively detect errors in the algorithms

C. Design a new value stream that ensures errors cannot be introduced

D. Reduce the number of algorithm changes so that IT staff have more time to spend on each one

Q: Which mid-level objective involves researching new products and services, as well as continuously evaluating existing products and services?

A. Valuable investments

B. Fast development

C. Resilient operations

D. Assured conformance

Q: In which circumstance would a transformation to high velocity IT be considered undesirable?

A. An organisation wants to support business enablement at speed

B. An organisation is used to innovating and operating quickly and effectively

C. An organisation consistently uses a high level of digital technology

D. An organisation believes that its values and leadership style are incompatible with the change

Course schedule

- ✓ Day 1: ITIL 4 Foundation
  - Key concepts, 4 dimensions of SM, ITIL Service Value System, the service value chain & 7 guiding principles
- ✓ Day 2: Create, Deliver and Support
  - How to use Value streams with support of ITIL practices
- ✓ Day 3: Drive Stakeholder Value
  - Customer journeys, relationships and service offerings
  - How to act together to ensure continual value co-creation
- ✓ Day 4: High Velocity IT
  - The Digital enterprise and how to use guiding principles
- Day 5: Deliver, Plan and Improve and Exam
  - Direction and Planning, GRC, Communication & OCM
  - Work with sample exam and examination

ITIL 4:
DIRECT, PLAN AND IMPROVE

Key learning requirements

DIRECT, PLAN & IMPROVE

- Understand the scope of what is to be directed and/or planned, and know how to use key principles and methods of direction and planning in that context
- Understand the role of GRC and know how to integrate the principles and methods into the service value system
- Understand and know how to use the key principles and methods of Communication and Organizational Change Management to direction planning and improvement

Introduction

Direct, Plan and Improve explores and explains principles, methods, and techniques that can be universally used to direct, plan and improve

Key terms defined and important concepts introduced here include:

- scope of control
- understand direction, vision, mission, and how to cascade objectives and requirements
- know how to define and use policy, guideline and controls
- understand how Governance, Risk and Compliance (GRC) impacts DPI
- how to use key principles and methods of communication and Organizational Change Management (OCM)

Strategy and strategic planning

Definition: Strategy
A broad approach or course of action defined by an organization for achieving its objectives

Strategies may be defined at multiple levels but, in organizations, strategies must cascade logically from the overall organizational strategy. They must also be tied to the achievement of objectives

Cascading objectives

When the organization's mission and strategy are understood, objectives can cascade from them, translating the strategy from one organizational level to the next

This ensures the organization's strategy, tactics, and operations are aligned and allows the reporting of accomplishments through feedback loops

Senior leaders can, therefore, monitor performance and make appropriate business decisions

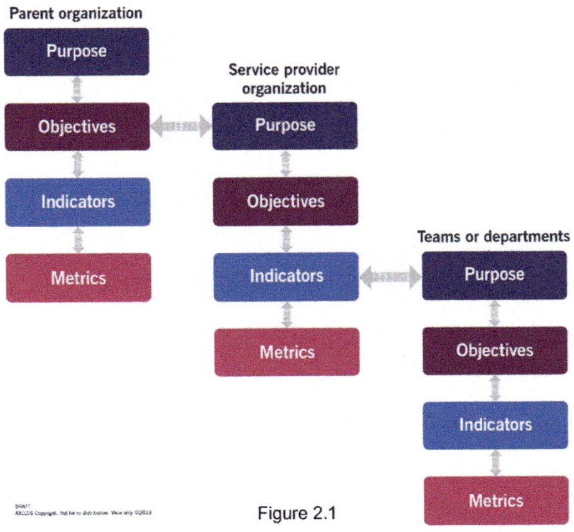

Figure 2.1

Cascading requirements

Just as objectives and plans cascade through an organization, so do the requirements for the resources (and capabilities) needed to achieve them

Requirements may relate to any of the four dimensions of service management

The achievement of a strategy relies on attention to all four dimensions of service management

Figure 2.2

Defining effective policies, controls, and guidelines

When defining policies, controls, and guidelines, it is important to remember why they are being defined in the first place. A policy that is defined but not followed is useless

An ineffective control is not a control at all. If those for whom a guideline is documented cannot use it, the effort to create the guideline was wasted

An organization's policies are part of its control landscape

Policy definition recommendations:

- *Be clear and concise*
- *Keep it simple and practical*
- *Anticipate questions*
- *Educate and communicate*
- *Build in flexibility*
- *Define the consequences of non-compliance*
- *Build in measurement and compliance validation*
- *Promote transparency*
- *Provide mechanisms for feedback*

Effective controls

When organizations implement controls, they aim to ensure that each control produces the desired result without creating unintended, undesirable consequences

Sometimes data entry systems are used to make relevant fields required to save a new record. However, placeholder values could be used leading to bad data and audit failure

Measurement and reporting are common controls. It is impractical, however, to measure everything. Measurements should be limited to those that can be actively used to make informed decisions. Regulatory controls cannot be marginalized

Automating controls or building them into technology relieves people of the effort of making the control work. Controls managed in this way should directly align with high-level objectives and should define what degree of variation or non-compliance, if any, is acceptable

What are sufficient controls?

How to ensure that controls are sufficient, but not excessive? When organizations implement controls, they aim to ensure that each control produces the desired result without creating unintended, undesirable consequences

Measurements should be limited to those that can be actively used to make informed decisions

Automating controls or building them into technology relieves people of the effort of making the control work

However, controls can fail or be circumvented. The organization must define, depending on the consequences, what degree of variation or non-compliance, if any, is acceptable for each control

Effective guidelines

Guidelines are not requirements, they are recommendations which allow for some discretion in their use. For guidelines to be effective, they must be easy to access, understand, and follow, and they must be truly useful.

Providing guidelines for every task in an organization is unnecessary. Situations in which guidelines might be useful include tasks that:

- are performed by many people, but that seem confusing without assistance
- used to be performed differently
- are infrequently performed but which should be performed consistently
- could be completed more easily or faster if organizational expertise was shared.

It is often helpful to ask for input when developing guidelines from the people closest to a task. Mechanisms for keeping guidelines current and correct are also crucial

Defining and ensuring compliance

Organizational governance ensures that:

- Stakeholder needs, conditions, and options are evaluated to determine balanced, established objectives
- Direction is set through prioritization and decision making
- Performance and compliance are monitored against established direction and objectives

Internal and external stakeholders help to ensure that organizations are compliant with rules and regulations and have the right governance systems in place to manage and sustain performance and conformity with compliance

The design and scope of the compliance mechanisms an organization implements will be strongly influenced by the regulatory environment

Some organizations find it useful to voluntarily conform to stricter regulations than needed

Role of risk management in DPI

Risk exists everywhere; it must be actively managed to ensure that the organization's objectives can be achieved

If direction is given without considering risks, the achievement of the associated objective will be suspect

Plans that do not note and mitigate risks are incomplete. Improvements are unlikely to succeed if the risks are unknown and unmanaged

Definition: Risk is a possible event that could cause harm or loss, or make it difficult to achieve objectives. Risk can also be defined as uncertainty of outcome and can be used in the context of measuring the probability of positive outcomes as well as negative outcomes

Role of risk in direction

The risk management practice should be continually applied and information about risks should be circulated when making decisions

Decisions about risks will vary depending on whether the risk relates to long-, medium-, or short-term organizational objectives

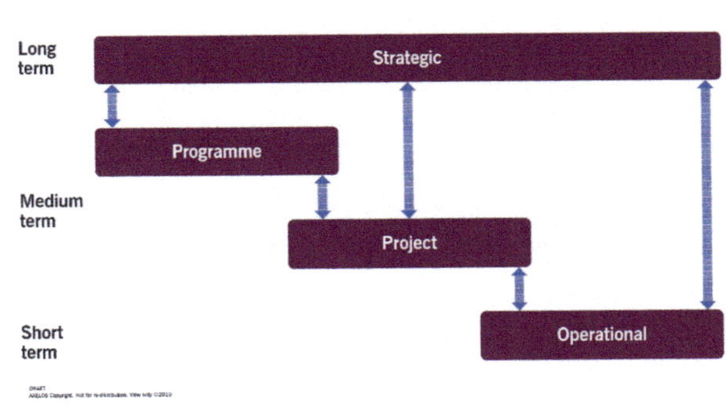

Figure 2.3 Interactions between long-, medium-, and short-term objectives

Everyone is responsible for participating in and contributing to risk management. Actively managing risks should be habitual and organisations should provide guidance, direction, and support so that people know which risks they can accept and which decisions must be escalated

Role of risk in planning & improvement

All plans must consider the associated risks and how those risks can be managed. If a plan's risks are too great, an alternative plan may need to be adopted

Every element of a plan must be considered and mitigation plans devised for risks that cannot be eliminated. Additionally, when a plan is underway, risks must be actively managed until the point of completion

These risk management requirements also apply to improvement activities. Improvements require change: change involves risk. When proposing action, the risks of proceeding should be considered, but so should the risks of inaction, which are equally relevant

Governance structures & roles

Decisions in organizations must align with the mission and strategy. Organizations use many different structures and methods for decision-making and directing activity and behaviours. It is critical for the organisation to place decision-making at the right level.

Governance structure	Role in governance
Board of directors	Responsible for their organization's governance. Specific responsibilities include: • setting strategic objectives • providing the leadership to implement strategy • supervising management • reporting to shareholders
Shareholders	Responsible for appointing directors and auditors to ensure effective governance
Audit committee	Responsible for supporting the board of directors by providing an independent assessment of management performance and conformance

Governance of the service provider

No single department can govern itself, unless that authority has been delegated by the governing body. When a service provider is part of a larger organization, the parent organization's governing body has authority over it. Some organizations delegate authority and perform certain governance activities at lower levels, based on that level's scope of control. In these cases, the organization's governing body oversees the delegated activities to ensure alignment with the organization's objectives

The ITIL SVS can be applied to an organization as a whole or to one or more of its departments. When the SVS is applied to a department, the governance component should be understood to come from the parent organization

Governing bodies do not function in a vacuum. They must consider external factors, such as regulations and legislation

Decision-making at the right level

Governance decisions are made at the highest levels of an organization, but most decisions should be made by other teams or areas in the organization. As much authority as possible should be delegated, so long as required outcomes are consistently produced.

When everyone has a defined role and knows their scope of control, they can make decisions within that scope and drive productive action. If their scope of control is too small, decisions will be forced upwards, slowing work and overloading decision-makers. People without the authority to make decisions may feel under-valued, which can lead to low productivity, an unwillingness to accept new responsibilities, and destructive behaviours.

Decisions that present little risk should be performed, as far as possible, by those performing the related work. A large proportion of decisions can be placed at the operational level when risk is moderated with training, automation, policies, and guidelines

Impact of governance on DPI

In the context of direction, planning, and improvement, governance's impacts are clear. Direction from a governing body will dictate the parameters of directions issued throughout the organization. If a direction given to an individual contradicts the broader direction from the governing body, the individual will struggle. Organizations must communicate about and oversee their governance decisions carefully and fairly

Governance decisions and directives are critical inputs to all planning. Plans must be made with a good understanding of what constitutes a successful outcome. When organizations issue new directives or defines new strategies, current plans should be reviewed

Finally, alignment with governance directives is an important element of every improvement opportunity. Every element of the SVS contributes to achieving strategic objectives and is therefore subject to continual improvement

Communication and Organisational Change Management (OCM)

OCM is concerned with the human side of change. It is a structured approach that ensures that improvements are implemented smoothly and successfully, facilitating lasting benefits

Improvements require people to change, which can include the way they work, their behaviour, or their roles. Regardless of the scope, size, or nature of an improvement initiative, there will be an impact on people. If people understand the change's purpose, how it will affect them, and believe in its importance and benefits, improvement initiatives are more likely to succeed

OCM aims to convince people of a change's value in order to reduce resistance and ensure the change is implemented and sustained successfully. Leading and implementing OCM requires a somewhat specialized skillset. Many organizations seek help from external suppliers, but the accountability for OCM cannot be transferred to an external resource

Communication and Organisational Change Management (OCM)

OCM is concerned with the human side of change. It is a structured approach that ensures that improvements are implemented smoothly and successfully, facilitating lasting benefits

Requiremen t	Details	How OCM helps
Clear and relevant objectives	To gain maximum support, improvements require objectives that are clear enough for people to understand and which make sense relative to the target organization	Improvement objectives must be communicated to stakeholders, who should then discuss them. If adjustments are made, the nature and reasons for them must also be communicated
Strong and committed leadership	It is critical that improvements are actively supported by leaders within the organization. If their commitments are visible overall buy-in will likely increase	Each sponsor and leader should be identified and their roles and responsibilities communicated to the initiative's stakeholders
Willing and prepared participants	People may resist a change for a variety of reasons. However, improvements need participants who are willing to change. People are often more willing to change when they feel they are suitably prepared	OCM allows for resistance to identified, understood, and overcome using a resistance management plan. OCM uses a training plan to ensure that people have the skills and knowledge to change successfully and a communication plan to manage change updates
Demonstrate d value	To keep moving forward, stakeholders must be convinced of the change's value before it is achieved and able to recognize the value after it has been achieved	OCM runs communication programmes to share any expected and achieved benefits with stakeholders, solidifying their commitment to the current effort and willingness to support future similar efforts
Sustained improvement	Many improvements fail when people revert to old ways of working. Even when an individual improvement succeeds, organizations can fail to sustain the momentum with more improvement	The OCM practice seeks to continually reinforce the value of the change through regular communication and the support of sponsors and leaders

OCM throughout DPI

The OCM practice should be woven throughout the work of direction, planning, and improvement.

It is almost impossible to be successful in any activity without considering the human factor; direction, planning, and improvement are no exception

OCM and direction

Previous sections have outlined the importance of establishing a vision and mission and using these to direct and drive behaviour

In this context, OCM is about addressing the human factor in relation to direction

If an organization wishes to direct their employees to behave in a certain way, perhaps to adopt a new way of working or use new criteria for their own decision-making, OCM principles and methods will help to uncover and overcome resistance to those directives

OCM and planning

When planning actions that involve significant change, OCM efforts should be integrated into the plan

Many plans are unsuccessful because people are not committed to them. OCM plans are typically defined within larger plans, particularly for programmes with multiple sub-projects and broad organizational impacts

However, the application of OCM is not only reserved for large programmes and projects. Even in routine planning at an operational level, individuals should consider the human factor and plan to address it accordingly

OCM and improvement

Without proper OCM, improvements cannot be achieved or sustained because improvement requires change and change requires people's participation and commitment.

The OCM practice is involved in improvement in two complimentary ways:

- It ensures that people involved in implementing improvement initiatives do so effectively and efficiently
- It ensures that people impacted by changes resulting from improvement initiatives accept and adopt those changes

Each stakeholder group's contributions to improvement at each level should be understood, and the most effective methods for communicating with them should be defined. Some stakeholders may need to be involved at a very detailed level, while others can be involved as reviewers or approvers. The identification and management of stakeholders is a core competency associated with OCM

Communication with stakeholders

The first step in defining a communications plan is to identify and understand the relevant stakeholders. Questions to ask include:

- Who are they?
- What interest do they have in the outcome of the work?
- What is their primary motivation?
- What information do they need?
- How do they want to communicate/receive information?
- What is their current opinion of the initiative?
- Are their opinions based on good information?
- Who, including other stakeholders, influences their opinions?
- If they not yet positive, what will cause them to support the improvement?
- If it is that they will support the improvement, how can their opposition be managed?
- Who, including other stakeholders, might be influenced by their opinions?

Stake-holders (group or individual)	Interest / involvement (How does the initiative impact the stakeholder?)	Power / influence / impact (on the initiative) (high / medium / low)	Expectations (What's in it for me?)	Observations / comments

Table 6.3 Sample stakeholder analysis worksheet

Communication with stakeholders

Once the stakeholders have been identified and characterized, each stakeholder's power, influence, and interest needs to be assessed. A stakeholder map can map stakeholders' levels of influence and involvement

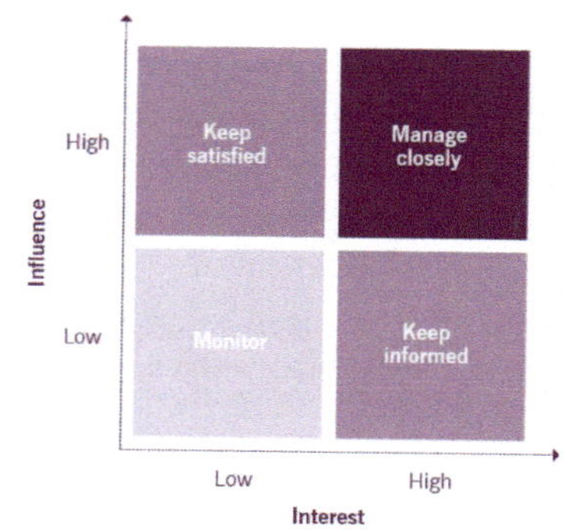

Figure 6.1 Sample stakeholder map

When identifying stakeholders, it is important to remember:

- Powerful stakeholders' opinions can shape initiatives at an early stage, which can improve the quality of the initiative and increases the likelihood that those stakeholders will demonstrate support later on
- Gaining support from powerful stakeholders can help win more resources, increasing the likelihood that the initiative will be successful
- Communicating with stakeholders early and frequently helps them to understand the nature and benefits of the initiative, which may encourage their vocal support
- Reactions to the initiative can be anticipated and actions that will win support can be planned in advance

Creating a communication plan

Once stakeholders have been identified and mapped, it is important to define a communication plan. This process follows six steps:

i. Planning the approach
ii. Define what is needed from each stakeholder
iii. Identify the message
iv. Devise a practical plan
v. Keep the initiative's best supporters engaged
vi. Consider how actions will affect the stakeholders

The value of good communication

Good communication is fundamental to almost every human endeavour. It is important to consider how organizations and people can develop and deploy communication to the best effect as part of their service management work

The value of good communication can be expressed from two perspectives: the costs stemming from its lack and the benefits gained through its achievement

Value streams, practices, and projects do not succeed without people collaborating and coordinating Poor communication can ruin good plans and can lead to waste. Good communication enables people to be informed, aware, and able to react quickly to potential issues

Communication principles

The fundamental principles required for good communication can be summarised as:

1. **Communication is a two-way process**
2. **We are all communicating, all the time**
3. **Timing and frequency matters**
4. **There is no single method of communicating that works for everyone**
5. **The message is in the medium**

Understanding, recognising and acting upon these principles is vital in order to build positive relationships with colleagues, customers and all stakeholders

1. Communication is a two-way process

- Communication is constant, multi-lateral, and pervasive. Listening and observing are key skills that have a big impact on successful collaboration
- The communicator cannot assume that their message has been acknowledged and understood. It is his responsibility to ensure that the message's purpose is understood and any actions needed undertaken
- The receiver also has a responsibility to confirm that their understanding of the message is correct
- To improve the quality and two-way nature of communications, research opinions and how people have responded to information sent to or expected from them
- People are more likely to engage in meaningful, two-way communication when they see that the other party is making a real effort to listen to their thoughts

2. We are all communicating all the time

- People constantly convey messages, purposefully or not, through their body language and tone of voice and, sometimes, by deliberately withholding these signals. Often, body language and tone have more of an impact than word choice
- Good communicators have high emotional intelligence, which allows them to interpret, understand, and predict people's actions and reactions. Communication requires an acknowledgement of the other person's emotional state and the use of flexible content and tone to achieve the desired objective
- At an operational level, the concept of setting the tone in a team is useful. It is important that the team is able to empathize with their customers and users, who are more likely to get in contact if they feel comfortable and cared for
- Everyone should understand how their communication style might impact how they are perceived by colleagues and stakeholders. Change agents should leverage a portfolio of communications techniques to get the best results

3. Timing and frequency matter

- In any successful relationship, good timing is essential. Communication must be proportionate, relevant, and appropriate at the time. Service providers can make mistakes when communicating, including:
- failing to respond to heightened activity and priority during key business periods
- sending updates about planned service downtime before it has been agreed with users and/or customers
- reporting great performance statistics when a major incident is occurring
- issuing satisfaction surveys before incidents are closed
- over-communicating to the point where messages lose their impact and get ignored
- To avoid these issues, consider the following:
- Develop relationships with service consumers and other stakeholders
- Understand the bigger picture
- Finally, parties should communicate regularly, to avoid poorly timed communications

4. There is no single method that works for everyone

- Everyone has different preferences for sending and receiving communication
- It is good practice to utilize several techniques to ensure that the intended audience is reached
- It is also good practice to tailor your communication to specific audiences, where possible
- It is also important to check that sent messages have been received and understood. Over time, organizations and individuals can observe patterns of effective communication methods based on their audiences and the communication's purpose
- It is useful to be aware of the ways people work and what their individual communication styles are
- When one method of communication proves ineffective or unpersuasive, it is essential to consider other options for the format and content of the message

5. The message is in the medium

- A message's format and nature often evokes an emotional response, which can drastically impact the level of importance, interest, and understanding that it will be afforded
- Therefore, it is important to select a message's format, style, size, and medium carefully to improve the likelihood of it being read and understood rather than ignored or deleted

Communication methods & media

- To ensure good communication it's important to choose the right method. Different communication methods that can be used:
- One on one / face to face
- Group meetings / workshops
- Telephone
- Email
- Instant messaging
- Social media
- Graphics
- By example
- Recordkeeping / workflow management applications
- Storytelling
- Reports

Feedback channels

- Communication is a two-way process; in addition to actively eliciting feedback, static feedback channels should be available and known to stakeholders. Multiple channels could be needed based on the type of feedback and the need for privacy or anonymity for those submitting it
- To share feedback, stakeholders must be secure and confident that it will receive attention. It is often easiest to leave the choice of anonymous or personalized feedback to the stakeholders
- Monitoring feedback on a communication approach is a vital way to ascertain whether messages are being received as intended, and it enables the refinement of future communications to better meet the needs of the audience

Ensure that all feedback gets a response or, the next time they are asked, stakeholders may be reluctant to provide it

Summary

We have just talked about:

- ✓ Understand the scope of what is to be directed and/or planned, and know how to use key principles and methods of direction and planning in that context
- ✓ Understand the role of GRC and know how to integrate the principles and methods into the service value system
- ✓ Understand and know how to use the key principles and methods of Communication and Organizational Change Management to direction planning and improvement

Q: A company is pursuing a strategy of acquiring small companies whose products complement their overall portfolio. Although early efforts to integrate company acquisitions have been successful, recent employee engagement results indicate a growing disengagement, especially among IT staff. IT staff feel that management do not listen or care about their concerns. What can be done to improve the engagement of IT staff with the changes?

A. Encourage the staff to suggest ways they can support the company strategy

B. Enable the staff to provide feedback through formal channels and create a dialogue with management

C. Arrange informal celebrations to acknowledge the work the staff have been doing

D. Establish a formal mentorship programme to connect IT staff from the company with the staff from acquired companies

Q: An improvement project failed to create the expected value because the team did not spent enough time thinking about how successful the solution would be. What would have helped to prevent this failure?

A. Kanban

B. Risk management

C. Measurement and reporting

D. Cascading objectives

Q: When implementing a change that impacts people, at which point should 'organisational change management' begin?

A. When project plans have been finalised

B. When planning changes to service components

C. When initiating a change programme

D. When planning the launch of a new product or service component

Q: The CIO of a large multi-national organisation has received complaints that work is slowing down as staff are often overwhelmed when trying to comply with controls. This is creating unintended and undesirable effects in other parts of the organisation. In response, the CIO has asked their management team to ensure that existing controls are sufficient, but not excessive. What should the management team NOT do when identifying excessive controls?

A. Work across multiple stakeholder groups to identify what data is useful for the organisation

B. Increase the amount of data measured and reported on, because the data might be needed in the future

C. Determine if staff are using placeholder data to complete mandatory fields, so they can continue to work if the required data is not available

D. Review external drivers, particularly legal obligations that impact the way the organisation works

Course schedule

- ✓ Day 1: ITIL 4 Foundation
  - Key concepts, 4 dimensions of SM, ITIL Service Value System, the service value chain & 7 guiding principles
- ✓ Day 2: Create, Deliver and Support
  - How to use Value streams with support of ITIL practices
- ✓ Day 3: Drive Stakeholder Value
  - Customer journeys, relationships and service offerings
  - How to act together to ensure continual value co-creation
- ✓ Day 4: High Velocity IT
  - The Digital enterprise and how to use guiding principles
- ✓ Day 5: Deliver, Plan and Improve and Exam
  - Direction and Planning, GRC, Communication & OCM
  - Work with sample exam and examination

Certification scheme

The ITIL 4 certification scheme that has two main training streams and three levels:

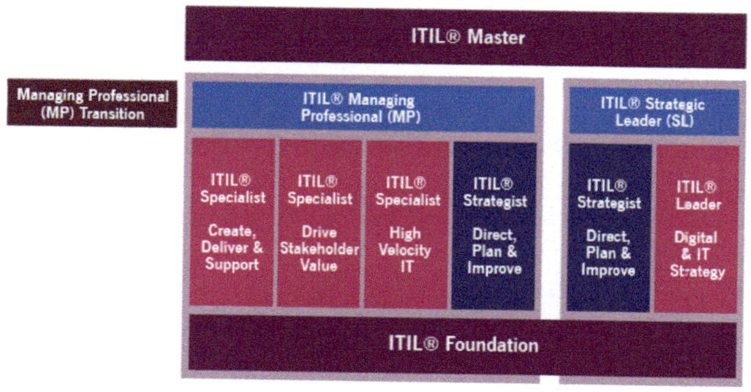

- Designation achieved once completed all relevant examinable modules in each stream
- Examinable modules towards ITIL Managing Professional and ITIL Strategic Leader
- Examinable module applicable to both ITIL Managing Professional and ITIL Strategic Leader
- Transition module for v3 ITIL Experts or those with 17 credits or more to gain ITIL Managing Professional designation

Examination preparation

The ITIL 4 Managing Professional Transition exam in short:

- 40 multiple-choice questions
- Each question is worth 1 mark. There are 40 marks available
- There is no negative marking
- Pass rate is 70% or higher – a raw score of 28 marks or above
- Duration is 90 minutes (113 minutes if not native in English)
- This is a 'closed book' examination. No materials other than the examination materials are permitted.

Good luck with the exam!

Thank you!

The ITIL® 4 Managing Professional Transition Examination

Sample Paper 1

Question Booklet

Multiple Choice

Examination Duration: 1 hour and 30 minutes

Instructions

1. You should attempt all 40 questions. Each question is worth one mark.
2. There is only one correct answer per question.
3. You need to answer 28 questions correctly to pass the exam.
4. Mark your answers on the answer sheet provided. Use a pencil (NOT pen).
5. You have 1 hour and 30 minutes to complete this exam.
6. This is a 'closed book' exam. No material other than the exam paper is allowed.

1) An organization has set up an 'ideas' page on an internal website and is encouraging its employees to experiment with different working methods when they experience issues that prevent the achievement of outcomes.

   Which concept is involved in these changes?

   A. Integration and data sharing
   B. Advanced analytics
   C. Team culture and differences
   D. The cu ture of continual improvement

2) Which is an example of team collaboration?

   A. Working with others to achieve individual goals
   B. Installing technology to facilitate communication among team members
   C. Working together to achieve a shared organizational objective
   D. Aligning the goals and KPIs of all individuals and groups

3) What does a digital organization have to do to create or maintain a competitive advantage?

   A. Market and sell existing products and services to new customers
   B. Deliver innovative new products and services quickly and reliably
   C. Provide multiple, integrated support channels for products and services
   D. Change the way that the IT function develops products and services

4) An organization has historically had a 'failure is not an option' culture. As a result, it is common for employees to hide mistakes.

Which would BEST help this organization foster a more open and trusting environment where people do not blame each other for failures?

A. Safety culture
B. Relationship management
C. Service-dominant logic
D. Cynefin

5) A stock trading company uses artificial intelligence (AI) to analyse markets and suggest investment strategies. The IT staff work long hours under pressure to manage the algorithms required. An increasing number of errors are occurring, and staff are disagreeing and blaming each other.

What approach should be taken to improve the situation?

A. Reinforce trust by fostering a safe environment in which staff can voice opinions and resolve issues
B. Introduce chaos engineering to proactively detect errors in the algorithms
C. Design a new value stream that ensures that errors cannot be introduced
D. Reduce the number of algorithm changes so that IT staff have more time to spend on each one

6) A service provider is launching a new service. The target market is users who have limited experience of using the internet and are unlikely to use social media.

Which is the BEST method of providing user support?

A. Provide simple on-line support and provide contact numbers for the service desk
B. Use machine learning chatbots to anticipate the needs of the users and provide solutions
C. Implement a 'shift left' approach to provide support and downloadable help articles
D. Use popular networking sites to promote and provide on-line user support

7) An organization that has a hierarchical structure is reviewing its project management practice and is considering the various ways to introduce a new or changed service.

Which is an advantage of an agile approach?

A. The ability to focus on increasing speed to market and making changes in line with business objectives
B. The option of using industry-standard project management models to deliver a phased, predictable project
C. The ability to focus on a clear release date which the various work streams can be aligned to
D. The ability to maintain its current structure which would speed up decision-making

8) Which is NOT a key focus of the 'information and technology' dimension?

A. Security and compliance
B. Communication systems and knowledge bases
C. Workflow management and inventory systems
D. Roles and responsibilities

9) Which describes user-centered service design?

A. Building a prototype of the minimum functionality that can be produced quickly
B. Applying the MoSCoW technique to a set of user requirements
C. Balancing user experience with the technical and business requirements
D. Using value stream mapping to identify a set of user requirements

10) What is an attribute of a strategic partnership?

A. Minimal information is shared
B. Trust needs to be developed
C. The customer needs to be able to exit easily
D. The relationship is driven by price

11) Which describes the nature of the guiding principles?

A. Guiding principles can guide an organization in all circumstances
B. Each guiding principle mandates specific actions and decisions
C. An organization will select and adopt only one of the seven guiding principles
D. Guiding principles describe the processes that all organizations must adopt

12) A company is pursuing a strategy of acquiring small companies which have products that complement their overall portfolio. Although early efforts to integrate company acquisitions have been successful, recent employee engagement results indicate a growing disengagement, especially among IT staff. IT staff feel that their managers do not listen or care about their concerns.

What can be done to improve the engagement of IT staff with the changes?

A. Encourage the staff to suggest ways they can support the company strategy
B. Enable the staff to provide feedback through formal channels and create a dialogue with management
C. Arrange informal celebrations to acknowledge the work the staff have been doing
D. Establish a formal mentorship programme to connect IT staff from the company with the staff from acquired companies

13) An improvement project failed to create the expected value because the team did not spend enough time thinking about how successful the solution would be.

What would have helped to prevent this failure?

A. Kanban
B. Risk management
C. Measurement and reporting
D. Cascading objectives

14) Which mid-level objective involves researching new products and services, and continuously evaluating existing products and services?

A. Valuable investments
B. Fast development
C. Resilient operations
D. Assured conformance

15) Which statement about the end-to-end customer journey is CORRECT?

A. It focuses on achieving outcomes
B. It represents a pre-determined path
C. It reflects an overall perception
D. It involves optimizing touchpoints

16) A service provider encourages their users to form a user support community.

What is a potential benefit to the service provider?

A. Reduction in demand for support from the service provider
B. Creation of groups that can be used to provision services
C. Improved collaboration within the service provider organization
D. Reduced collaboration between user organizations

17) Which are methods of measuring customer perception of IT services?

1. Service availability reports
2. Feedback from service review meetings
3. Telephone surveys
4. Service provider employee satisfaction surveys

A. 1 and 2
B. 2 and 3
C. 3 and 4
D. 1 and 4

18) An organization is experiencing delays to incident resolutions because there is a lack of clarity about escalation paths among support teams. These delays sometimes cause the organization to lose significant amounts of money. The organization has decided to investigate the activities involved in incident resolution, and to produce a flow of activities which are connected from the time a disruption occurs until a resolution has been identified.

Which concept is this an example of?

A. Organizational structure
B. Collaboration
C. Value stream
D. Workforce planning

19) Which concept is concerned with creating good working relationships with other people by considering their intellectual and emotional needs?

A. Employee satisfaction measurement
B. The value of positive communications
C. Organizational structure
D. Automated interactions

20) In which circumstance would a transformation to high velocity IT be considered undesirable?

A. An organization wants to support business enablement at speed
B. An organization is used to innovating and operating quickly and effectively
C. An organization consistently uses a high level of digital technology
D. An organization believes that its values and leadership style are incompatible with the change

21) A service provider has built a 'cooperative relationship' with a customer.

Which activity are they MOST LIKELY to use to validate the services that are provided?

A. Ad-hoc joint service reviews of costs and benefits
B. Continual tracking and analysis of the outcomes, costs, and risks
C. Joint service reviews of achievements of service targets
D. Review of costs of service provider technology upgrades

22) When implementing a change that impacts people, at which point should 'organizational change management' begin?

A. When project plans have been finalized
B. When planning changes to service components
C. When initiating a change programme
D. When planning the launch of a new product or service component

23) The CIO of a large multinational organization has received complaints that work is slowing down as staff are often overwhelmed when trying to comply with controls. This is creating unintended and undesirable effects in other parts of the organization. In response, the CIO has asked their management team to ensure that existing controls are sufficient, but not excessive.

What should the management team NOT do when identifying excessive controls?

A. Work across multiple stakeholder groups to identify what data is useful for the organization
B. Increase the amount of data measured and reported on because the data might be needed in the future
C. Determine if staff are using placeholder data to complete mandatory fields so they can continue to work if the required data is not available
D. Review external drivers, particularly legal obligations that impact the way the organization works

24) Which high velocity IT objective applies to all the service value chain activities?

A. Fast development
B. Valuable investments
C. Assured conformance
D. Resilient operations

25) What should be carried out FIRST when applying the 'focus on value' guiding principle?

A. Identifying the outcomes that the service facilitates
B. Identifying all the suppliers and partners involved in the service
C. Determining who the service consumer is in each situation
D. Determining the cost of providing the service

26) A small service provider is experiencing growth and success. Currently all important decisions are made by a small executive group. This creates delays because members of the group are often unavailable.

Which is the BEST approach for establishing an authority structure for decision-making within the service provider organization?

A. Appoint one responsible member of the executive group to make all important decisions
B. Ensure that technical decisions are made by the operational staff who can define the risks
C. Keep high risk decisions within the executive group, but define a policy for delegating other decisions
D. Refer all decisions to the executive group, who will consider each case and delegate when appropriate

27) An organization is adopting agile methodologies. The IT department is centralized at the corporate office. The infrastructure operations manager has become very frustrated with the development teams using agile because they are not following any of the established standards.

What is the BEST method of communication to improve the situation between the operations manager and the agile development teams?

A. Use email communication to ensure that there is a clear record of all exchanges
B. Arrange a phone call between the operations manager and the teams using agile
C. Organize a face-to-face meeting between the operations manager and the teams using agile
D. Use the organization's social media page to explain why agile is the best approach

28) An organization uses internal and external development teams that collaborate well to produce new and upgraded services. When the changes are released, users complain that the service desk staff are often not aware of the changes and that any support issues relating to the changes take a long time to resolve.

What would prevent this situation from occurring in the future?

A. Outsource the organization's internal support provision to the external development organization
B. Introduce a self-service system for users so that they do not have to rely on the support teams
C. Engage, during the development phase, the support staff who are involved in the value stream for the service
D. Allow users direct access to second-line support teams, bypassing the service desk

29) An organization intends to improve the quality of support of its services. It recognizes that some ways of working are not focused on creating value.

Which is an example of a working practice which the organization should STOP?

A. Discussing requirements and expectations of service availability with customers to determine incident prioritization categories
B. Conducting a survey to collect feedback from users on how they would like the interface of the self-service portal to look
C. Engaging with consumers to understand if the service support model will be fit for purpose following a planned organizational re-structure
D. Using the response and fix times from the organization's hardware vendor to drive incident resolutions times for customers

30) A business analyst has recognized that business managers have conflicting hypotheses about their needs for departmental improvement. They also find it difficult to articulate their requirements.

Which approach is BEST for obtaining requirements?

A. Conduct an interview based on a carefully prepared script of detailed questions
B. Allocate a week to analyse how the department works, and then make logical conclusions
C. Develop multiple prototypes of a possible solution, and adopt what works best
D. Ask the director to enforce what the business analyst considers to be the most plausible solution

31) Which concept is concerned with creating prototypes to test hypotheses?

A. Design thinking
B. Safety culture
C. Continual improvement
D. Complexity thinking

32) What is a potential benefit of marketing services to internal customers?

A. It reduces the need to include services in the service catalogue
B. It utilizes services more effectively and efficiently
C. It maximizes the profit of the service provider
D. It reduces the demand for services

33) Which ITIL concept describes governance?

A. The seven guiding principles
B. The four dimensions of service management
C. The service value chain
D. The service value system

34) A management consultant has been hired to help a multinational organization improve corporate governance, risk management, and compliance.

Who would the consultant PRIMARILY work with to understand the current state of governance?

A. The board of directors
B. The largest shareholders
C. The internal audit committee
D. The service management office

35) An organization is struggling to respond to and resolve incidents in accordance with business needs and expectations. High priority incidents are often ignored because low priority incidents can be easier and quicker to solve.

A. A triage system
B. A CI/CD pipeline
C. Service integration and management
D. Deep learning

36) A service provider describes a package that includes a laptop with software, licenses, and support.

What is this package an example of?

A. Value
B. An outcome
C. Warranty
D. A service offering

37) What is a challenge when onboarding individual consumers?

A. Ensuring that the sponsor agrees the level of service that the consumer receives
B. Dealing with many service consumers with different skills
C. Returning consumer equipment and cancelling user rights
D. Identifying and documenting service requirements

38) A network security engineer is defining the controls needed to protect firewalls.

What is the FIRST thing they should do to determine the correct controls?

A. Review existing firewall rules and controls to ensure that there has been no deterioration
B. Review the risk register to ensure that all risks are being managed
C. Contact the firewall vendor to identify commonly used controls for this technology
D. Identify the organization's objectives that the controls need to support

39) Which value chain activity ensures people understand the organization's vision?

A. Improve
B. Plan
C. Deliver and support
D. Obtain/build

40) A retail organization is trying to improve customer loyalty. One approach it is taking is to allow its online shopping customers to choose the time of day that their package is delivered to their home.

What is this an example of?

A. Continuous delivery
B. Service-dominant logic
C. Deployment management
D. Availability management

END OF EXAMINATION

The ITIL® 4 Managing Professional Transition Examination

Sample Paper 1

Answers and Rationales

For exam paper: EN_ITIL4_TRANS_2019_SamplePaper1_QuestionBk_v1.1

Q	A	Syllabus Ref	Rationale
1	D	CDS 1.3.b	A. Incorrect. Integration and data sharing focuses on bringing together multiple systems within service design. Ref 3.1 B. Incorrect. "Advanced analytics is the autonomous or semi-autonomous examination of data or content using high level techniques and tools that go beyond traditional business intelligence, to discover profounder insights, make predictions, or generate recommendations." Ref 3.2 C. Incorrect. A team culture is made up "of values that are shared by a group of people, including their ideas, beliefs, and practices, and their expectations with regard to how the individuals within the group should behave." Ref 2.3.1 D. Correct. These changes are examples of the 'building trust' element of a continual improvement culture. This element is focused on establishing a comfort zone where people feel enabled and supported in trying out new ideas, making suggestions and experimenting. "It is essential for all stakeholders to understand the importance of positive attitude, collaboration, transparent working, and a supportive culture. This should encourage individuals to make suggestions, regardless of how unusual the suggestion may seem, so long as the goal is to improve the service." Ref 2.3.4

Q	A	Syllabus Ref	Rationale
2	C	CDS 1.1.b	A. Incorrect. This is an example of cooperation. Cooperation is working with others to achieve your goals, which may be part of a common goal. With cooperation "There is a risk that individuals or teams who are cooperating instead work in silos. As a result, the individual or team goals are achieved but the organizational goals are missed." Ref 2.3.5 B. Incorrect. Collaboration is more than just technology, i.e., Skype, Slack, or MS Office Teams. Collaboration is "Working with others to achieve common shared goals.". "From a business perspective, collaboration is a practice where individuals work together to achieve a common, shared goal/objective." Ref 2.3.5 C. Correct. From a business perspective, "collaboration is a practice where individuals work together to achieve a common, shared goal/objective". Ref 2.3.5 D. Incorrect. Aligning the goals and KPIs of all the individuals and groups is not sufficient for collaboration, these need to be shared and integrated, for cooperation to become collaboration. A clear understanding of how all of the individuals and groups are successful and their value in contributing to the success is necessary for effective collaboration. Ref 2.3.5
3	B	HVIT 1.1.a	A. Incorrect. "An organization which is enabled by digital technology to do business that it could not do in a same or similar way, or at a similar scale, without this technology." Marketing and selling existing products and services is not a different way to do business. Ref 2.2 B. Correct. "For digital products and customer experiences, innovative digital investments are needed to create or maintain competitive advantage, and have to be realized quickly. The resulting digital products, services and customer channels have to be operationally resilient, and the users of these product must use them well to achieve the desired return on investment." Ref 2.3 C. Incorrect. This is a description of IT transformation. "IT transformation is focused on how IS and IT services are developed, run and supported." Ref 2.4.1 D. Incorrect. This is a description of IT transformation. "IT transformation is often used to denote major change which improves how IT services are provided." Ref 2.4.1

Q	A	Syllabus Ref	Rationale
4	A	HVIT 3.1.b	A. Correct. Safety culture involves creating a climate in which people are comfortable being (and expressing) themselves, even in stressful situations such as in the event of a failure. "Be realistic about failure: acknowledge that it will happen and that people are not to blame, but the system". Ref 3.3.2.2, B. Incorrect. 'Relationship management' relates to how an organization establishes and maintains its relationships with stakeholders. Activities and resources associated with this technique are "Activities and techniques to regularly review the status and direction of existing relationships with various stakeholders". Ref 4.2.3 tab 4.7 C. Incorrect. Service-dominant relates to a service provider involving a customer in the service delivery process to co-create value. "Service-dominant logic regards service as the process of doing something for and with another party. Value creation is a collaborative process. In service-dominant logic, value is always co-created." Ref 2.5.2.5 D. Incorrect. Cynefin is a decision-making framework. "The Cynefin sense-making framework... offers a practical way of assessing complexity and determining appropriate courses of action." Ref 3.2.3.1
5	A	HVIT 3.2.b	A. Correct. "In HVIT environments, it is crucial that people feel able to share their opinions and experiment with improvement without the fear of judgement or Ref 3.2.2.2 B. Incorrect. Chaos engineering, through tools such as the Chaos Monkey "tests the resilience of IT infrastructures and platforms". This would not detect errors in financial trading algorithms. Ref 4.3.2 C. Incorrect. It is not possible to remove all errors in a high velocity IT environment. "These systems always contain multiple flaws and therefore latent issues. Continual changes to the system and its environment mean that the flaws also continually change." Ref 3.2.2.2 D. Incorrect. This organization has invested in AI in order to co-create value. The IT staff need to support this capability. "In the digital organization, IT drives and enables the business." Ref 4.1

Q	A	Syllabus Ref	Rationale
6	A	DSV 6.6	A. Correct. Providing a combination of online and service desk support is the most appropriate for the target market described in the question. Ref 7.3, tab 7.7 B. Incorrect. Chatbots have "limited applicability" and would not be much help in providing support to the described target market. Also, this solution does not suggest backing up chatbots with human support. "Don't replace human interfaces with ML-based until level of success is high enough – provide human backup". Ref 7.3, tab 7.7 C. Incorrect. A 'shift left' approach is only useful when users are sufficiently experienced with using the internet. In this case we are told this is not true. Challenges of 'shift left' include, "Users do not have sufficient technology skills and/or motivation to use the self-help tools; Only limited range of tasks can be fulfilled by users at their level of access to the service; Mistakes made by users during self-help may cause more incidents; Knowledge base navigation may be difficult". Ref 7.3, tab 7.7 D. Incorrect. This answer suggests using social media which the target market is stated to have little experience of, hence it is inappropriate. Social media "is widely used for services provided to individual users that are likely to be active users of social networks". Ref 7.3

Q	A	Syllabus Ref	Rationale
7	A	CDS 2.1	A. Correct. An agile approach encourages "small increments that provide fast feedback and the opportunity to change specifications at short notice". It would therefore provide the advantage of increasing speed to market and making changes in line with business objectives. Ref 4.2.1 B. Incorrect. This is a waterfall approach, which "defines the process of system development/implementation as a linear series of phases, with each phase only commenced once the previous step has been completed." Ref 3.7.1 C. Incorrect. A waterfall approach, rather than an agile approach, has the advantage of having a go live date as a clear target for all involved to focus and co-ordinate activities. "Plan-based approaches, such as the waterfall method, may still be more suitable in some situations; for example, where there is a high certainty about the requirements of the service", for example when it will be delivered. Ref 3.7.5 D. Incorrect. The organization's structure would not work well with an agile way of working, and this would not speed up decision-making. "A hierarchical organizational structure can impede decision-making, as well as an organization's ability to work quickly and with an agile approach". Ref 2.4
8	D	FND 3.1.b	A. Incorrect. "The challenges of information management, such as those presented by security and regulatory compliance requirements" are a focus of the 'information and technology' dimension". Ref 3.2 B. Incorrect. Key focuses of the 'information and technology' dimension "include, but are not limited to, workflow management systems, knowledge bases, inventory systems, communication systems, and analytical tools". Ref 3.2 C. Incorrect. Key focuses of the 'information and technology' dimension "include, but are not limited to, workflow management systems, knowledge bases, inventory systems, communication systems, and analytical tools." Ref 3.2 D. Correct. "The organizations and people dimension of a service covers roles and responsibilities, formal organizational structures, culture, and required staffing and competencies, all of which are related to the creation, delivery, and improvement of a service." This is not a key focus of the 'information and technology' dimension. Ref 3.1

Q	A	Syllabus Ref	Rationale
9	C	DSV 4.1	A. Incorrect. This refers to the technique of identifying the minimum viable product (MVP). " A minimum viable product is a product with just enough features to satisfy early customers and to provide feedback for future product development." It is not a user-centered method of designing services. Ref 5.2.4 B. Incorrect. “The MoSCoW method is a prioritization technique for managing requirements.” MOSCOW is used for specifying requirements, not for designing service offerings. Ref 5.2.67 C. Correct. "User-centered design ensures that the product and the service focus on what users need before balancing this with the technical and the user experience." Ref 5.3.3 D. Incorrect. “Value stream mapping is Lean technique for illustrating and analysing the logic of a value stream.” VSM is used to identify areas for improvement. It is not a user-centered method of designing services. Ref 5.3.1
10	B	DSV 3.2	A. Incorrect. An attribute of a "basic" relationship is that there is not much data-sharing. This is not true of a "partnership" relationship. Ref 4.2, tab 4.3 B. Correct. The need to develop a deep sense of trust is an attribute of a strategic partnership. Ref 4.2, tab 4.3 C. Incorrect. The need for the service consumer to exit easily is an attribute of a "basic" relationship. This is not true of a "partnership" relationship. Ref 4.2, tab 4.3 D. Incorrect. An attribute of a "basic" relationship is that it is "driven by price". This is not true of a "partnership" relationship. Ref 4.2, tab 4.3

Q	A	Syllabus Ref	Rationale
11	A	FND 2.1	A. Correct. A guiding principle is defined as a recommendation that can guide an organization in all circumstances and will guide organizations when adopting service management. They are not described as prescriptive or mandatory. Ref 4.3 B. Incorrect. The guiding principles will be reviewed and adopted by organizations. The guiding principles guide organizations to make decisions and adopt actions. They do not mandate specific actions and decisions. Ref 4.3.8 C. Incorrect. Organizations will use the principles relevant to them and are not mandated to use a given number. Ref 4.3 D. Incorrect. The guiding principles guide organizations to make decisions and adopt actions. They are not mandatory. Ref 4.3

Q	A	Syllabus Ref	Rationale
12	B	DPI 5.2.c	A. Incorrect. This answer focuses on encouraging staff members to provide feedback, but does not address the relevant challenge of staff feeling management is not listening nor caring about their concerns. "Communication is a two-way process; in addition to actively eliciting feedback, static feedback channels should be available and known to stakeholders." This necessitates establishing formal feedback channels where staff feel safe to voice their concerns and suggestions. For their part, management must respond to all feedback to encourage continued feedback and engagement. Ref 6.1.6 B. Correct. This answer focuses on establishing formal feedback channels where staff feel safe to voice their concerns and suggestions. This will open up channels to listen to staff concerns. It is important, however for management to respond to all feedback to encourage continued feedback and engagement. "Communication is a two-way process; in addition to actively eliciting feedback, static feedback channels should be available and known to stakeholders." Ref 6.1.6 C. Incorrect. In this answer, management is taking action to recognize and celebrate the work and success of the teams, which may prove helpful, but does not address the relevant challenge of staff feeling that management is not listening nor caring about their concerns. In fact, this approach can produce negative results as it may come across as insincere. It is important to establish formal feedback channels where staff feel safe to voice their concerns and suggestions. When feedback is given, management must respond to encourage continued feedback and engagement. Ref 6.1.6 D. Incorrect. This answer focuses on bridging the culture between company staff and the staff at acquired companies. While this may prove helpful, it does not address the relevant challenge of staff feeling management is not listening nor caring about their feedback. For that, it is important to establish formal feedback channels where staff feel safe to voice their concerns and suggestions. For their part, management must respond to all feedback to encourage continued feedback and engagement. Ref 6.1.6

Q	A	Syllabus Ref	Rationale
13	B	DPI 3.1	A. Incorrect. The use of a Kanban board provides "full visualization of the process workflow", it is not suitable for evaluating how well future solutions would work. Ref 7.3.3.7 B. Correct. "If risks are not properly understood, teams could be directed to undertake projects that are certain to fail. If team members notice risks, but do not see evidence of preventative actions, they may lose confidence in their project, increasing the likelihood of failure." Ref 1.2.1.3 C. Incorrect. Measurement and reporting might help identify an issue, but it would not help to predict issues in advance. When measuring and reporting "data is collected, processed, and reported on to validate that desired actions are being performed or that agreed objectives are being met." Ref 2.5.2.1 D. Incorrect. "When the organization's mission and strategy are understood, objectives can cascade from them, translating the strategy from one organizational level to the next. This ensures the organization's strategy, tactics, and operations are aligned". This will not help the team to think about how well a proposed solution will work. Ref 2.1.3
14	A	HVIT 1.3.a	A. Correct. The valuable investments goal involves identifying and justifying digital investments. "New digital products and services should be envisaged and evaluated in terms of profitability", and "It is also important to continually evaluate investments after they have been justified and approved, because more valuable options for investment may exist. " Ref 4.1 B. Incorrect. Fast development involves realizing "new and improved digital products and services frequently, quickly and reliably." Ref 4.2 C. Incorrect. Resilient operations involves ensuring "that digital products are available for use whenever needed." Ref 4.3 D. Incorrect. Assured conformance involves assuring "that service provision and service consumption comply with corporate and regulatory directives with respect to governance, risk and compliance." There is no reference to corporate and regulatory directives in this question. Ref 4.5

Q	A	Syllabus Ref	Rationale
15	C	DSV 1.1	A. Incorrect. This is just one perspective. "Mastering the customer journey helps to maximize stakeholder value through co-creation, focusing on both outcomes and experience." Ref 2 B. Incorrect. A customer journey may start at different points and follow different paths. "The customer journey rarely follows a predefined path between touchpoints and service interactions". Ref 2.2 C. Correct. The customer journey represents "the overall perception of the touchpoints and interactions between the service consumer and the service provider… Every customer journey involves several touchpoints and service interactions between the service provider, service consumer, and other stakeholders." Ref 1.1.4, 2.2 D. Incorrect. The customer journey includes touchpoint and interactions and the customer's experience relative to each. "Individual touchpoints may perform well even if the overall experience is poor. Customers experience service providers through end-to-end experiences, not touchpoints." Ref 2.2
16	A	DSV 7.3	A. Correct. "Peer support, knowledge articles, discussions, and improvement initiatives can help to decrease users' demand for support from the service provider". Ref 8.3 B. Incorrect. User communities are valuable for peer support, discussions and improvements, but do not provide sufficient formality for identifying which users are entitled to use what services. "Peer support, knowledge articles, discussions, and improvement initiatives can help to decrease users' demand for support from the service provider". Ref 8.3 C. Incorrect. While the creation of user communities may lead to improved collaboration within the user community, and between the users and the service provider, it is unlikely to impact the level of collaboration with the service provider organization. "Communities can become a powerful means of communication and cooperation between users and service providers." Ref 8.3 D. Incorrect. The creation of user communities might provide increased collaboration between user organizations. "Communities can become a powerful means of communication and cooperation between users and service providers." Ref 8.3

Q	A	Syllabus Ref	Rationale
17	B	DSV 8.1	B. Correct. (2) Feedback from service review meetings is a good method of understanding customer perception of the service. "The service provider should consider using the following methods to monitor customer experience: ...gathering customer feedback from service review meetings". Ref 9.2.2 (3) Feedback from telephone surveys is a good method of understanding customer perception of the service. "The service provider should consider using the following methods to monitor customer experience: ...conducting telephone perception surveys". Ref 9.2.2 A. C. D. Incorrect. (1) Service availability reports do not provide information about the perception of the service. Service availability targets may be met but customer perception can be low. "The service provider should consider using the following methods to monitor customer experience: ...gathering customer feedback from service review meetings". Ref 9.2.2 (4) Service provider employee satisfaction surveys will only measure service provider perception of the service, not customer perception. Ref 9.3
18	C	CDS 2.3	A. Incorrect. Organizational structure is concerned with how to build and maintain different structures in organizations. Ref 2.1.1 B. Incorrect. Collaboration is a " Working with others to achieve common shared goals." "From a business perspective, collaboration is a practice where individuals work together to achieve a shared goal or objective." Ref 2.3.5 C. Correct. A value stream is a "series of steps an organization undertakes to create and deliver products and services to consumers." In this example, the value stream is focused on reducing value leakage. Ref 4.1.1 D. Incorrect. The focus of workforce planning is on enabling "organizations, leaders, and managers to focus on creating an effective and actionable people strategy". Ref 2.2.3

Q	A	Syllabus Ref	Rationale
19	B	CDS 1.1.g	A. Incorrect. "Therefore, organizations should measure employee satisfaction in order to understand how well they are meeting the employees' changing needs and expectations." Ref 2.2.4 B. Correct. "Good human communication is about being efficient, responsive, professional, and effective. Effective human communication is enhanced by establishing positive relationships that avoid unnecessary issues and stress." Ref 2.3.7 C. Incorrect. Organizational structure is concerned with the approaches to building and managing team structures. Ref 2.1.1 D. Incorrect. The Automation of interactions does not address the intellectual and emotional needs of others as effectively as positive communications. "Most working projects, teams, initiatives, and organizations require productive and positive interactions between individuals to succeed. Human interaction and communication are where real people still stand apart, ahead of the machines." Ref 2.3.7
20	D	HVIT 1.2	A. Incorrect. "High velocity IT is the application of digital technology for significant business enablement, where time to market, time to customer, time to change, and speed in general is crucial." Ref 2.7 B. Incorrect. "High velocity does not come at the expense of the utility or warranty of the solution, so high velocity equates with high performance in general", and "In the context of high velocity IT, this is interpreted as doing the right thing – in other words, not only fulfilling requirements but making the right decisions about investment and sustainability." Ref 2.1 C. Incorrect. Digital technology is "technology that digitizes something, or processes digital data. Digitization is the process of transforming something from analogue form to digital form, in other words expressed in (binary) digits, or processing digital data." The use of digital technology would support the move to high velocity IT. Ref 2.2, tab 2.2 D. Correct. Organizations "may choose not to pursue an increase in velocity as the amount of cultural change involved is deemed to be too difficult, or be unlikely to generate an acceptable return on investment." Ref 2.1

Q	A	Syllabus Ref	Rationale
21	C	DSV 8.5	A. Incorrect. Ad-hoc reviews are appropriate for a "basic relationship" but not a "cooperative relationship". Shared activities for a basic relationship include "Ad-hoc service review". Ref 9.1, tab 9.2 B. Incorrect. Continual tracking and analysis of the outcomes, costs and risks is appropriate for a "partnership relationship" but not a "cooperative relationship". Shared activities for a partnership relationship include "Continual tracking and analysis of the outcomes, costs and risks and seeking optimization". Ref 9.1, tab 9.2 C. Correct. Shared activities for a cooperative relationship include "Joint service review of achievements compared to agreements and promises". Ref 9.1, tab 9.2 D. Incorrect. A review of costs of technology upgrades is a concern of the service provider only and not a joint or collaborative activity. Further the review will not validate the services that are provided. Ref 9.1, tab 9.2
22	C	DPI 5.1	A. Incorrect. Organizational change management should begin as soon as planning begins and "woven throughout the work of direction, planning, and improvement." Ref 6.3.2 B. Incorrect. Organizational change management should begin as soon as planning begins and "woven throughout the work of direction, planning, and improvement." Ref 6.3.2 C. Correct. Organizational change management should begin as soon as planning begins and "woven throughout the work of direction, planning, and improvement." Ref 6.3.2 D. Incorrect. Organizational change management should begin as soon as planning begins and "woven throughout the work of direction, planning, and improvement." Ref 6.3.2

Q	A	Syllabus Ref	Rationale
23	B	DPI 3.3	A. Incorrect. The management team should promote collaboration and transparency between the organization and its consumers to build trust and validate the usefulness of measurements, which are "common controls". Data that is not valuable might indicate that excessive control is behind the unnecessary measurement and reporting of that data. "Measurements should be limited to those that can be actively used to make informed decisions." Ref 2.5.2.1 B. Correct. "It is impractical, however, to measure everything. Measurements should be limited to those that can be actively used to make informed decisions." The management team should avoid doing so, and use the guiding principle 'focus on value' to identify the data needed to make better decisions. Ref 2.5.2.1 C. Incorrect. The management team should pay attention to unintentional consequences as a way of identifying controls that are excessive. A control to enforce mandatory fields on records can inadvertently lead to staff using placeholder data to be able to save the record and continue working. Ref 2.5.2.1 D. Incorrect. The management team should review external factors, "when designing organizational measures, it is important to account for external factors, particularly legal factors, which are often enforced by regulatory authorities and should be considered mandatory." Ref 2.5.2.1
24	C	HVIT 2.1.c	A. Incorrect. "Fast development is mainly achieved by the application development and infrastructure engineering that takes place in the design and transition and obtain/build value chain activities." Ref 2.5.1 B. Incorrect. "Valuable investments is mainly achieved by the decision-making that occurs in the plan value chain activity". Ref 2.5.1 C. Correct. The five objectives are closely related to specific value chain activities, except for assured conformance, that applies to the whole value chain. Ref 2.5.1 D. Incorrect. "Resilient operations is mainly achieved by running the system in the delivery and support value chain activity". Ref 2.5.1

Q	A	Syllabus Ref	Rationale
25	C	FND 2.2.a	A. Incorrect. It is essential to determine who the service consumer is and what they value. The outcomes should be based on this understanding, rather than determining them. "The first step in focusing on value is knowing who is being served. In each situation the service provider must, therefore, determine who the service consumer is". Ref 4.3.1.1 B. Incorrect. Suppliers and partners are possible stakeholders, but it is important to identify the service consumer first. "The first step in focusing on value is knowing who is being served. In each situation the service provider must, therefore, determine who the service consumer is". Ref 4.3.1.1 C. Correct. "The first step in focusing on value is knowing who is being served. In each situation the service provider must, therefore, determine who the service consumer is". Ref 4.3.1.1 D. Incorrect. The cost of providing the service may have some impact on the value from the perspective of the service provider. But "The first step in focusing on value is knowing who is being served. In each situation the service provider must, therefore, determine who the service consumer is". Ref 4.3.1.1

Q	A	Syllabus Ref	Rationale
26	C	DPI 2.1.c	A. Incorrect. Appointing one member of the executive group does not address the availability of the executive group. "Governance decisions are made at the highest levels of an organization, but most decisions should be made by other teams or areas in the organization. As much authority as possible should be delegated, so long as required outcomes are consistently produced." Ref 2.2.2 B. Incorrect. Operational staff may not be aware of all risks, technical and business. "Governance decisions are made at the highest levels of an organization, but most decisions should be made by other teams or areas in the organization. As much authority as possible should be delegated, so long as required outcomes are consistently produced." Ref 2.2.2 C. Correct. The decision-making authority should be aligned with risk and with business objectives. The 'risk management' practice should be continually applied and information about risks should be circulated when making decisions. Decisions about risks will vary depending on whether the risk relates to long-, medium-, or short-term organizational objectives, and "Governance decisions are made at the highest levels of an organization, but most decisions should be made by other teams or areas in the organization. As much authority as possible should be delegated, so long as required outcomes are consistently produced." Ref 2.2.2, 2.3.1 D. Incorrect. Referring all decisions to the executive group does not address the availability of the executive group. "Governance decisions are made at the highest levels of an organization, but most decisions should be made by other teams or areas in the organization. As much authority as possible should be delegated, so long as required outcomes are consistently produced." Ref 2.2.2

Q	A	Syllabus Ref	Rationale
27	C	DPI 5.2.b	A. Incorrect. "Email is particularly useful for short, factual information exchanges, especially where written documentation is helpful", so it would not be the best method of communication for building relationships. Ref tab 6.2 B. Incorrect. While "It is possible to infer certain aspects of a person's emotional state while using a telephone through their tone of voice, volume, and language choices," a face-to-face meeting is more likely to resolve conflict. Ref tab 6.2 C. Correct. Face-to-face communication is the best approach where there is need to build the relationship and resolve issues. "One-on-one interactions are the best way of having good interactions and build relationships. They should always be considered when resolving issues." Ref tab 6.2 D. Incorrect. While social media can be useful for internal communication, "It requires a good understand of the technologies of use," and "Many organizations have policies governing social media's use." Face-to-face communication would be better because it does not come with these risks. Ref tab 6.2

Q	A	Syllabus Ref	Rationale
28	C	CDS 2.1	A. Incorrect. The problem faced by this organization is that the relevant stakeholders have not been included in the development phase. There is nothing to suggest that the external development organization's support teams would have been involved, if the internal support teams were not. A change to the sourcing arrangements will not necessarily make any difference to the user experience. Ref 4.2.2.2 B. Incorrect. This answer does not resolve the situation in which the relevant stakeholders have not been involved in the development phase. Even if a self-service system were to be introduced, such a system would not be able to deal with all user issues relating to the changes, and the support staff would still need to be involved at some stage. Ref 4.2.2.2 C. Correct. "Successful completion of this step requires the organization to engage with the requestor, as well as other stakeholders like marketing and sample users (using surveys and polls) to fill in the business case template with high-level information about the requirement, benefits (both quantitative and qualitative), costs, and risks. This information is also supplemented by high-level estimates from various technical and service management teams, who consider the cost of development, release, operations and support." Ref 4.2.2.2 D. Incorrect. There is nothing to suggest that the second-line support teams are any more aware of the changes going live than the service desk staff are. But, even if the second-line support teams are able to resolve users' queries, the service desk, as first-line support, should not be bypassed, and should be made aware of the changes. Ref 4.2.2.2

Q	A	Syllabus Ref	Rationale
29	D	CDS 2.3	A. Incorrect. Discussing with customers their requirements and expectations of service availability in order to determine incident prioritization categories is an activity that adds value for the customer and the organization. This would help organization to identify the requirements and expectation of service availability and how the issues are prioritized and managed to support business continuity. The data can also be used to identify actual levels of performance for future improvements. This is a customer-orientated, 'outside in' approach, and should be maintained. "It is highly desirable to maintain an outside-in tone or language when documenting the value stream, for example, by... framing outcomes and value from the customer or user's point of view." Ref 4.1.3 B. Incorrect. This is a consumer-orientated, 'outside in' approach, which will help the organization to make the self-service portal more user-friendly and improving the customer experience. This is an activity that adds value for the customer and should be maintained. "It is highly desirable to maintain an outside-in tone or language when documenting the value stream, for example, by... framing outcomes and value from the customer or user's point of view." Ref 4.1.3 C. Incorrect. This is another example of 'outside in' approach, which will help the organization to decide whether or not their support model needs to be re-designed as per the planned organizational re-structure. This is an activity that adds value for the customer by ensuring customer requirements and expectations are considered for the service support model in the planned organizational structure. "It is highly desirable to maintain an outside-in tone or language when documenting the value stream, for example, by... framing outcomes and value from the customer or user's point of view." Ref 4.1.3 D. Correct. This is not an example of an 'outside in' approach, as the organization is using a supplier-orientated method to determine customer-related goals. In this option the organization is viewing requirements from the perspective of their own technical and functional structures and working according to their capability. To be really effective however, it is vital to start with the customer journey and business objective and then work backwards to reflect how to deliver as per business objectives and requirements. "It is highly desirable to maintain an outside-in tone or language when documenting the value stream, for example, by... framing outcomes and value from the customer or user's point of view." Ref 4.3.1

Q	A	Syllabus Ref	Rationale
30	C	HVIT 3.2.d	A. Incorrect. This approach is more suited to a highly predictable situation. "Obvious – clear causality, where predetermined best practice should be applied... apply constructs such as prescriptive processes and detailed waterfall-based plans". Ref 3.2.3.1 B. Incorrect. This approach is more suited when there is enough information to be analysed. "Complicated – unclear but knowable causality that can be determined by analysis or expertise, followed by good practice... apply constructs such as case management and timeboxing". Ref 3.2.3.1 C. Correct. This approach is best. "Complex – unclear and unknowable causality requiring safe-to-fail experimentation (emergent practice)". Ref 3.2.3.1 D. Incorrect. This approach is more suited to extremely urgent and important issues. "Chaotic – a more extreme form of complexity that demands immediate action to transition the situation to complex (novel practice)". Ref 3.2.3.1
31	A	HVIT 3.1.a	A. Correct. A typical behaviour pattern of 'design thinking' is to "create hypotheses based on observation and reflection, and test them with prototypes ." Ref 3.2.1.2 B. Incorrect. Safety culture is "a climate in which people are comfortable being (and expressing) themselves." Ref 4.5.2 C. Incorrect. The continual improvement model provides organizations with "a structured approach to implementing improvements." Ref 3.2.3 D. Incorrect. Complexity-based heuristics focusses on "The understanding that some systems are unpredictable and can only be influenced by experimentation with attractors, repellers, and modifications to the boundaries that constrain the system's parts." Ref 3.2.3

Q	A	Syllabus Ref	Rationale
32	B	DSV 4.2	A. Incorrect. Services should be included in the service catalogue regardless of whether they are used by internal or external customers. The service catalogue is a valuable tool for marketing services to internal customers. "One of the most important tools for the internal sales process is the service catalogue". Ref 5.4.2 B. Correct. "Internal sales and promotions, combined with incentives and pricing mechanisms, are important for managing demand. "Some benefits of selling to internal customers are: ...enhanced utilization of existing services". Ref 5.4.2 C. Incorrect. Gaining maximum profit should not be the goal of marketing and selling services. "The service provider should avoid creating a higher price for a service than it is worth for the customer." Many service providers do not charge for internal services and act as cost centres, not profit centres. Ref 5.4.3 D. Incorrect. Services should not be marketed to reduce demand. The appropriate marketing of services should increase demand. Ref 5.4.2
33	D	FND 4.1	A. Incorrect. The seven guiding principles are 'focus on value', 'start where you are', 'progress iteratively with feedback', 'collaborate and promote visibility', 'think and work holistically', 'keep it simple and practical' and 'optimize and automate'. Ref 4.3 B. Incorrect. The four dimensions of service management are 'organizations and people', 'information and technology', 'partners and suppliers', and 'value streams and processes'. Ref 3.1-3.4 C. Incorrect. The activities of the service value chain are 'plan', 'improve', 'engage', 'design and transition', 'obtain/build', and 'deliver and support'. Ref 4.5 D. Correct. The components of the service value system are 'guiding principles', 'governance', 'service value chain', 'practices', and 'continual improvement'. Ref 4.1

Q	A	Syllabus Ref	Rationale
34	A	DPI 3.2	A. Correct. The board of directors are "Responsible for their organization's governance". Ref 2.2.1, tab 2.1 B. Incorrect. Shareholders are "Responsible for appointing directors and auditors to ensure effective governance." Ref 2.2.1, tab 2 C. Incorrect. The audit committee is "Responsible for supporting the board of directors by providing an independent assessment of management performance and conformance." Ref 2.2.1, tab 2.1 D. Incorrect. A service management office is "A group or department that functions as a centre of excellence for service management, ensuring continual development and the consistent application of management practices across the organization." Ref 7.2.2
35	A	CDS 3.1.b	A. Correct. Triage is used to assess which issues are the most urgent. Triage prioritization is "determined by urgency, based upon an assessment of the impact a delay may cause." Ref 5.1.3 B. Incorrect. A CI/CD pipeline "defines the set of tools, integrations, practices, and guardrails which allow a continuous and substantially automated flow of changes, from their initial design and development, through to deployment into production." Ref 3.7.3 C. Incorrect. Service integration and management refers to an approach whereby organizations manage and integrate multiple suppliers in a value stream. Ref 5.2.4 D. Correct. Incorrect. "Deep learning is a subset of machine learning based on artificial neural networks." Ref 3.6.2

Q	A	Syllabus Ref	Rationale
36	D	FND 1.3.a	A. Incorrect. The combination of things described in this option may help to create value, but it is not an example of value. Value is "the perceived benefits, usefulness and importance of something." Ref 2.1 B. Incorrect. The combination of things described in this option may help to create an outcome, but it is not an example of an outcome. Outcome is "a result for a stakeholder enabled by one or more outputs." Ref 2.5.1 C. Incorrect. Warranty is "assurance that a product or service will meet agreed requirements." New functionality may or may not affect warranty. Ref 2.5.4 D. Correct. Service providers define combinations of goods, access to resources and service actions, to address the needs of different consumer groups. These combinations are called service offerings. Ref 2.3.2
37	B	DSV 6.2	A. Incorrect. It is the customer that agrees the level of service, not the sponsor. However, in the case of an individual consumer one person can take on all three roles of user, customer and sponsor. Therefore, agreeing the level of service is no more of a challenge than in any other consumer arrangement. Also, requirements are agreed during the agree step, not during onboarding. "Onboarding occurs after an agreement is reached or changed but before service consumption starts." Ref 7 B. Correct. Individual consumers refer to the case in which one individual performs all three roles of user, customer and sponsor. The challenges of fostering relationships with individual consumers include the "Large number of service consumers with different skills and backgrounds". Ref 7.2.2 tab 7.6 C. Incorrect. Returning equipment and revoking rights are typical activities of offboarding, not onboarding. Ref 7.6.1 D. Incorrect. Requirements are established during the offer step, not during onboarding. Onboarding takes place after the agreement is reached about starting service relationship or expanding the scope of existing relationship, and before service consumption starts. Ref 7

Q	A	Syllabus Ref	Rationale
38	D	DPI 2.1.b	A. Incorrect. This may be needed at a later stage, but the existing rules may have been based on organizational objectives that are no longer relevant or appropriate. "Automating controls or building them into technology relieves people of the effort of making the control work. Controls managed in this way should directly align with and support the achievement of high-level objectives." Ref 2.5.2.1 B. Incorrect. Controls are "means of managing a risk", and the risk register is one possible source of information about risks that need to be managed. But firstly the network engineer must understand the organizational objectives they are trying to achieve, otherwise they may implement controls that are not appropriate for the organization. "Automating controls or building them into technology relieves people of the effort of making the control work. Controls managed in this way should directly align with and support the achievement of high-level objectives." Ref 2.5.2.1, 1.2.1 C. Incorrect. It may be appropriate to talk to the vendor about what controls are available and how they are commonly used. But firstly, the network engineer must understand the organizational objectives they are trying to achieve, otherwise they may implement controls that are not appropriate for the organization. "Automating controls or building them into technology relieves people of the effort of making the control work. Controls managed in this way should directly align with and support the achievement of high-level objectives." Ref 2.5.2.1 D. Correct. "Automating controls or building them into technology relieves people of the effort of making the control work. Controls managed in this way should directly align with and support the achievement of high-level objectives." Ref 2.5.2.1

Q	A	Syllabus Ref	Rationale
39	B	FND 5.2.a	A. Incorrect. The purpose of the 'improve' value chain activity is "to ensure continual improvement of products, services, and practices across all value chain activities and the four dimensions of service management." Ref 4.5.2 B. Correct. The purpose of the 'plan' value chain activity is "to ensure a shared understanding of the vision, current status, and improvement direction for all four dimensions and all products and services across the organization." Ref 4.5.1 C. Incorrect. The purpose of the 'deliver and support' value chain activity is "to ensure that services are delivered and supported according to agreed specifications and stakeholders' expectations." Ref 4.5.6 D. Incorrect. The purpose of the 'obtain/build' value chain activity is "to ensure that service components are available when and where they are needed, and meet agreed specifications." Ref 4.5.5
40	B	HVIT 3.2.a	A. Incorrect. Continuous delivery "allow smaller, more frequent and more reliable deployments" Ref 3.2.2.3.3 B. Correct. Service-dominant logic involves the co-creating value and states that "value is always co-created", and "When service-dominant logic is applied to service delivery, the provider pays more attention to the customer's specific situation, and involves the customer in the service delivery process." Ref 2.5.2.5 C. Incorrect. The 'deployment management' practice involves moving components (e.g., hardware, software) to live environments. "Systematic/automatic installation of specific versions or packages of software to a predetermined environment (e.g. integration, UAT, production)." Ref 4.2.5 tab 4.10 D. Incorrect. The 'availability management' practice involves ensuring IT services deliver agreed levels of availability. "Detailed warranty requirements for the new or changed service negotiated and agreed with stakeholders." Ref 4.3.3 tab 4.16

ITIL®

ITIL® 4 Managing Professional

Transition Syllabus

AXELOS.com

AXELOS
GLOBAL BEST PRACTICE

October 2019

Contents

1 Introduction

The purpose of this document is to outline:

- the learning outcomes of the ITIL 4 Managing Professional Transition qualification and the assessment criteria that a candidate is expected to meet for each learning outcome (with reference to the ITIL 4 core publications and the *ITIL Practice Guide* library)
- the examination design, in terms of question types to be used, exam duration and administrative considerations
- the weightings (number of questions) across learning outcomes, assessment criteria and 'Bloom's level' (level of cognitive processing required to answer the question/task, according to Bloom's (revised) taxonomy).

The target audience for this document is:

- our Examination Institute partner, PeopleCert
- Accredited Organization partners.

The purpose of the ITIL 4 Managing Professional Transition Qualification is:

- to allow candidates of the previous iteration of ITIL the opportunity of a straightforward transition to ITIL 4 in order to achieve the designation of ITIL 4 Managing Professional
- to provide candidates with an understanding of the new ITIL 4 Foundation concepts and definitions, including the key differences between the previous iteration of ITIL and ITIL 4 and how they can be practically applied
- to provide candidates with an understanding of the key elements from each one of the four ITIL Managing Professional (MP) modules: Create, Deliver and Support, Drive Stakeholder Value, High Velocity IT and Direct, Plan and Improve, and also ITIL 4 Foundation.

The purpose of the ITIL 4 Managing Professional Transition Examination is:

- to assess whether the candidate can demonstrate sufficient understanding and practical application of the concepts covered in the following ITIL 4 Core publications to be awarded the designation ITIL 4 Managing Professional:
    - **ITIL 4 Foundation:** to introduce readers to the management of modern IT-enabled services, to provide them with an understanding of the common language and key concepts
    - **Create, Deliver and Support (CDS):** to provide the candidate with an understanding on how to integrate different value streams and activities to create, deliver and support IT-enabled products and services, and relevant practices, methods and tools
    - **Drive Stakeholder Value (DSV):** to provide the candidate with an understanding of all types of engagement and interactions between a service provider and their customers, users, suppliers and partners, including key CX, UX and journey mapping concepts
    - **High Velocity IT (HVIT):** to provide the candidate with an understanding of the ways in which digital organizations and digital operating models function in high velocity environments
    - **Direct, Plan and Improve (DPI):** to provide the candidate with the practical skills necessary to create a 'learning and improving' IT organization, with a strong and effective strategic direction.

The target audience for this qualification is:

- existing ITIL Expert qualification holders, and those who have 17 Credits with the ITIL credit scheme, wishing to develop their knowledge and application.

2 ITIL 4 Transition Syllabus

The table below specifies the learning outcomes of the ITIL 4 Transition qualification, and the assessment criteria used to assess a candidate's achievement of these learning outcomes, subsequent to a course of study. Note: Principal book references are in parentheses. These refer to the section, but not the subsections within it. The verb for each assessment criterion indicates the Bloom's level (BL): 'Recall'/'Define' indicates Level 1 basic recall and recognition, 'Describe'/'Understand', indicates Level 2 understanding/comprehension, and 'Know how to' indicates Level 3 application. This syllabus is subdivided into each core module from which the content was drawn

2.1 ITIL 4 FOUNDATION

Learning Outcome	Assessment Criteria	BL	No. marks
1. Understand the key concepts of service management	1.3 Describe the key concepts of service relationships: (2.3.2, 2.4, 2.4.1): a) Service offering b) Service relationship management c) Service provision d) Service consumption	BL2	1
2. Understand how the ITIL guiding principles can help an organization adopt and adapt service management	2.1 Describe the nature, use and interaction of the guiding principles (4.3, 4.3.8)	BL2	1
	2.2 Explain the use of the guiding principles (4.3): a) Focus on value (4.3.1 - 4.3.1.4) b) Start where you are (4.3.2 - 4.3.2.3) c) Progress iteratively with feedback (4.3.3 - 4.3.3.3) d) Collaborate and promote visibility (4.3.4 - 4.3.4.4) e) Think and work holistically (4.3.5 - 4.3.5.1) f) Keep it simple and practical (4.3.6 - 4.3.6.3) g) Optimize and automate (4.3.7 - 4.3.7.3)	BL2	1
3. Understand the four dimensions of service management	3.1 Describe the four dimensions of service management (3): a) Organizations and people (3.1) b) Information and technology (3.2) c) Partners and suppliers (3.3) d) Value streams and processes (3.4-3.4.2)	BL2	1
4. Understand the purpose and components of the ITIL service value system	4.1 Describe the ITIL service value system (4.1)	BL2	1
5. Understand the activities of the service value chain, and how they interconnect	5.1 Describe the interconnected nature of the service value chain and how this supports value streams (4.5)	BL2	1

Learning Outcome	Assessment Criteria	BL	No. marks
	5.2 Describe the purpose of each value chain activity: a) Plan (4.5.1) b) Improve (4.5.2) c) Engage (4.5.3) d) Design & transition (4.5.4) e) Obtain/build (4.5.5) f) Deliver & support (4.5.6)	BL2	

2.2 CREATE, DELIVER AND SUPPORT

Learning Outcome	Assessment Criteria	BL	No. marks
1. Understand how to plan and build a service value stream to create, deliver, and support services	1.1 Understand the concepts and challenges relating to the following across the service value system: a) Organisational structure (2.1.1) b) Collaborative culture (2.3.5,2.3.5.1-3) c) Teams, roles & competencies (2.2.1-2, 2.2.2.1-2) d) Team culture and differences (2.3.1-5) e) Working to a customer-orientated mindset (2.3.6, 2.3.6.1) f) Employee satisfaction management (2.2.4) g) The value of positive communications (2.3.7, 2.3.7.1)	BL2	2
	1.3 Know how to plan and manage resources in the service value system: a) Team collaboration and integration (2.3.3, 2.3.5) b) Workforce planning (2.2.3) c) Results based measuring and reporting (2.2.5, 2.2.5.1)	BL3	1
2. Know how relevant ITIL practices contribute to the creation, delivery and support across the SVS and value streams	2.1. Know how to use a value stream to design, develop and transition new services (4, 4.1, 4.1.1-7, 4.2.1-2, 4.2.2.1-4.2.2.7)	BL3	2
	2.3 Know how to use a value stream to provide user support (4, 4.1, 4.1.1-7, 4.2.1, 4.2.2, 4.2.2.1-8)	BL3	2
3. Know how to create, deliver and support services	3.1 Know how to co-ordinate, prioritize and structure work and activities to create deliver and support services, including: a) Managing queues and backlogs (5.1.1, 5.2.1, 5.2.1.1-2) b) Prioritizing work (5.1 including all subsections)	BL3	1

2.3 DRIVE STAKEHOLDER VALUE

Learning Outcome	Assessment Criteria	BL	No. marks
1. Understand how customer journeys are designed	1.1 Understand the concept of the customer journey (2, 2.1, 2.2)	BL2	1
3. Know how to foster stakeholder relationship	3.1 Understand the concepts mutual readiness and maturity (4.3.5)	BL2	1
	3.2 Understand the different supplier and partner relationship types, and how these are managed (4.4, 4.2, 4.2.1-3)	BL2	
	3.3 Know how to develop customer relationships (4.1, 4.1.1-2, 4.3, 4.3.1-4 (including all subsections)	BL3	
4. Know how to shape demand and define service offerings	4.1 Understand methods for designing digital service experiences based on value driven, data driven and user centred service design (5.3, 5.3.1-6)	BL2	1
	4.2 Understand approaches for selling and procuring service offerings (5.4, 5.4.1-3)	BL2	1
6. Know how to onboard and offboard customers and users	6.1 Understand key transition, onboarding and offboarding activities (7)	BL2	1
	6.2 Understand the ways of relating with users and fostering user relationships (7.2, 7.2.1-2)		
	6.3 Understand how users are authorized and entitled to services (7.4)		
	6.4 Understand different approaches to mutual elevation of customer, user and service provider capabilities (7.5)		
	6.5 Know how to prepare onboarding and offboarding plans (7.1, 7.1.1-4, 7.6, 7.6.1-2, 7.7)	BL3	1
	6.6 Know how to develop user engagement and delivery channels (7.3)		
7. Know how to act together to ensure continual value co-creation (service consumption / provisioning)	7.1 Understand how users can request services (8.2, 8.2.1-2, 8.2.4-5)	BL2	1
	7.3 Understand the concepts of customer and user (8.3, 8.3.1)		
	7.4 Understand methods for encouraging and managing customer and user feedback (8.2.6)		
	7.5 Understand the concept of 'moments of truth' (8.2.4)		

Learning Outcome	Assessment Criteria	BL	No. marks
8. Know how to realise and validate service value	8.1 Understand methods for measuring service usage and customer and user experience and satisfaction (9.2.2, 9.3.3)	BL2	1
	8.2 Understand methods to track and monitor service value (outcome, risk, cost and resources) (9.2, 9.2.1-3, 9.3)		
	8.3 Understand different types of reporting of service outcome and performance (9.2.1, 9.3, 9.5.1)		
	8.4 Understand charging mechanisms (9.4, 9.5.4)		
	8.5 Know how to validate service value (9.3)	BL3	1
	8.6 Know how to evaluate and improve the customer journey (9.4, 9.4.1-2)		

2.4 HIGH VELOCITY IT

Learning Outcome	Assessment Criteria	BL	No. marks
1. Understand concepts regarding the high-velocity nature of the digital enterprise, including the demand it places on IT	1.1 Understand the following terms: a) Digital organization (2.3) b) High velocity IT (2.1) c) Digital transformation (2.4) d) IT transformation (2.4.1) e) Digital product (2.6.1) f) Digital technology (2.2)	BL2	1
	1.2 Understand when the transformation to high velocity IT is desirable and feasible (2.1)	BL2	1
	1.3 Understand the five objectives associated with digital products – to achieve: a) Valuable investments – strategically innovative and effective application of IT (2.5.1, 4, 4.1) b) Fast development - quick realization and delivery of IT services and IT-related products (2.5.1, 4, 4.2) c) Resilient operations - highly resilient IT services and IT-related products (2.5.1, 4, 4.3) d) Co-created value - effective interactions between service provider and consumer (2.5.1, 4, 4.4) e) Assured conformance - to governance, risk and compliance (GRC) requirements (2.5.1, 4, 4.5)	BL2	1
2. Understand the digital product lifecycle in terms of the ITIL 'operating model'	2.1 Understand how high velocity IT relates to: a) The four dimensions of service management (2.6.6, 2.6.6.1-4, 2.6.7)	BL2	1

Learning Outcome	Assessment Criteria	BL	No. marks
	b) The ITIL service value system (2.6, 2.6.3, 2.6.3.1, 2.6.3.2, 2.6.4) c) The service value chain (2.5.1, 2.6.3, 2.6.3.1, 2.6.3.2, 2.6.4) d) The digital product lifecycle (2.6.2)		
3. Understand the importance of the ITIL Guiding Principles and other fundamental concepts for delivering high velocity IT	3.1 Understand the following principles, models and concepts: a) Ethics (3.2.1, 3.2.1.1) b) Safety culture (3.2.2.2, tab 3.2) c) Lean culture (3.2.3.2, tab 3.3) d) Toyota Kata (3.2.3.3.1) e) Lean / Agile / Resilient / Continuous (2.5.2, 2.5.2.1-4, tab 2.2) f) Service-dominant logic (2.5.2.5) g) Design thinking (3.2.1.2) h) Complexity thinking (3.2.3.1)	BL2	2
	3.2 Know how to use the following principles, models and concepts: - Ethics - Safety culture - Lean culture - Toyota Kata - Lean / Agile / Resilient / Continuous - Service-dominant logic - Design thinking - Complexity thinking (3.2, 3.2.1, 3.2.1.1, 3.2.2.2, tab 3.2, 3.2.3.2, tab 3.3, 3.2.3.3.1, 2.5.2, 2.5.2.1-4, tab 2.2, 2.5.2.5, 3.2.1.2, 3.2.3.1) to contribute to: a) Help get customers' jobs done (3.1.4, tab 3.1) b) Trust and be trusted (3.1.2, tab 3.1) c) Continually raise the bar (3.1.3, tab 3.1) d) Accept ambiguity and uncertainty (3.1.1, tab 3.1) e) Commit to continual learning (3.1.5, tab 3.1)	BL3	3

2.5 DIRECT, PLAN & IMPROVE

Learning Outcome	Assessment Criteria	BL	No. marks
2. Understand the scope of what is to be directed and/or planned, and know how to use key principles and methods of direction and planning in that context	2.1 Identify the scope of control and within this: a) Know how to cascade objectives and requirements (1.3.1.1, 2.1.3, 2.1.4) b) Know how to define effective policies, controls and guidelines (2.5.2, 2.5.2.1-2, 2.5.3) c) Know how to place decision-making authority at the correct level (2.2.2)	BL3	2
3. Understand the role of GRC and know how to integrate the principles and methods into the service value system	3.1 Understand the role of risk and risk management in DPI (2.3, 2.3.1-2)	BL2	1
	3.2 Understand how governance impacts DPI (2.2, 2.2.1, 2.2.1.1, 2.2.2, 2.2.3)	BL2	1
	3.3 Know how to ensure that controls are sufficient, but not excessive (2.5.2.1)	BL3	1
5. Understand and know how to use the key principles and methods of Communication and Organizational Change Management to direction planning and improvement	5.1 Understand the nature, scope and potential benefits of organizational change management (6.3, 6.3.1, 6.3.2, 6.3.2.1-3, the 'organizational change management' practice)	BL2	1
	5.2 Know how to use the key principles and methods of Communication & OCM a) Identify and manage different types of stakeholders (6.2, 6.2.1, 6.2.2) b) Effectively communicate with and influence others (6.1.2 - 6.1.2.5, 6.3.2, 6.3.2.1-3, tab 6.2) c) Establish effective feedback channels (6.1.2.1, 6.1.5, 6.1.6)	BL3	2

3 ITIL 4 Transition Examination Design

3.1 EXAMINATION ADMINISTRATION

Duration: 90 minutes

NOTE: Candidates taking the exam in a language that is not their native or working language may be awarded 25% extra time, i.e. 113 minutes in total.

Materials permitted: This is a 'closed book' examination. No materials other than the examination materials are permitted.

Prerequisites: The candidate must have the ITIL Expert certification or have 17 credits under the ITIL Credit scheme. In addition, the candidate must have attended an accredited training course for this module (the recommended duration for this training is 30 hours including the examination).

3.2 QUESTION TYPES

All 40 questions are Objective Test Questions (OTQs), which present four options from which one option is selected. Distractors (wrong answers) are options that candidates with incomplete knowledge or skill would be likely to choose. These are generally plausible responses relating to the syllabus area being examined. Question styles used within this type are: 'standard', 'list' (2 correct items), and, exceptionally, 'negative' standard OTQ.

<u>Example 'standard' OTQ:</u>

Which is a source of best practice?

a) Q
b) P
c) R
d) S

<u>Example 'list' OTQ:</u>

Which statement about service asset and configuration management is CORRECT?

1. It does Q
2. It does P
3. It does R
4. It does S

a) 1 and 2
b) 2 and 3
c) 3 and 4
d) 1 and 4

NOTE: Two of the list items are correct. List style questions are never negative.

<u>Example 'negative' standard OTQ:</u>

Which is NOT a defined area of value?

a) Q
b) P
c) R
d) S

NOTE: Negative questions are <u>only used as an exception</u>, where part of the learning outcome is to know that something is not done or should not occur.

3.3 SCORING

Number of questions: 40

Marks: Each question is worth 1 mark. There are 40 marks available.

There is no negative marking.

Provisional pass mark: 70% or higher - a raw score of 28 marks or above

3.4 WEIGHTINGS BY BLOOM'S LEVEL

There are 23 questions at Bloom's Level 2 = approx. 57.5%

There are 17 questions at Bloom's Level 3 = approx. 42.5%

3.5 WEIGHTINGS BY LEARNING OUTCOME

Module	Learning Outcome	No.OTQs	Approx. weighting
Foundation	1. Understand the key concepts of service management	1	2.5%
	2. Understand how the ITIL guiding principles can help an organization adopt and adapt service management	2	5%
	3. Understand the four dimensions of service management	1	2.5%
	4. Understand the purpose and components of the ITIL service value system	1	2.5%
	5. Understand the activities of the service value chain, and how they interconnect	1	2.5%
CDS	1. Understand how to plan and build a service value stream to create, deliver, and support services	3	7.5%
	2. Know how relevant ITIL practices contribute to the creation, delivery and support across the SVS and Value streams	4	10%
	3. Know how to create, deliver and support services	1	2.5%
DSV	1. Understand how customer journeys are designed	1	2.5%

Module	Learning Outcome	No.OTQs	Approx. weighting
	3. Know how to foster stakeholder relationships	1	2.5%
	4. Know how to shape demand and define service offerings	2	5%
	6. Know how to onboard and offboard customers and users	2	5%
	7. Know how to act together to ensure continual value co-creation (service consumption / provisioning)	1	2.5%
	8. Know how to realise and validate service value	2	5%
HVIT	1. Understand concepts regarding the high-velocity nature of the digital enterprise, including the demand it places on IT	3	7.5%
	2. Understand the digital product lifecycle in terms of the ITIL 'operating model'	1	2.5%
	3. Understand the importance of the ITIL Guiding Principles and other fundamental concepts for delivering high velocity IT	5	12.5%
DPI	2. Understand the scope of what is to be directed and/or planned, and know how to use key principles and methods of direction and planning in that context	2	5%
	3. Understand the role of GRC and know how to integrate the principles and methods into the service value system	3	7.5%
	5. Understand and know how to use the key principles and methods of Communication and Organizational Change Management to direction, planning and improvement	3	7.5%

COMMERCIAL IN CONFIDENCE